Green Energy and Technology

For further volumes:
http://www.springer.com/series/8059

Hermann-Josef Wagner
Jyotirmay Mathur

Introduction to Hydro Energy Systems

Basics, Technology and Operation

Springer

Prof. Dr.-Ing Hermann-Josef Wagner
Lehrstuhl Energiesysteme und
Energiewirtschaft
Universität Bochum
Universtitätsstr. 150
44801 Bochum
Germany
e-mail: lee@lee.rub.de

Dr.-Ing Jyotirmay Mathur
Department of Mechanical Engineering
Malaviya National Institute of Technology
Jawahar Lal Nehru Marg
302017 Jaipur
India
e-mail: jyotirmay.mathur@gmail.com

ISSN 1865-3529 e-ISSN 1865-3537
ISBN 978-3-642-20708-2 e-ISBN 978-3-642-20709-9
DOI 10.1007/978-3-642-20709-9
Springer Heidelberg Dordrecht London New York

Cover design: WMXDesign GmbH

Printed on acid-free paper

Springer is part of Springer Science+Business Media (www.springer.com)

Preface

Water is one of the essential elements of life. Besides this it is also utilized as sustainable source of energy, worldwide. The power generation technology using water has been developed and has matured over the time to permit an economic harnessing of the power contained in water. Due to a wide variety of settings for power generation, the technical design and operation of hydropower plants vary over a wide range and are influenced by local requirements and practices. The power generation capacity ranges from a few kilowatts to several thousand megawatts, and locations are found from the remotest water streams to large river systems.

This book explains different variants of hydropower plants. The authors have tried to strike a balance between a short book chapter and a very detailed book for subject experts. There are several primary reasons for doing so: first, the field of hydro power is quite inter-disciplinary and requires simplified presentation for a person from a non-parent discipline. The second reason is related to students and engineers who are starting out in the hydropower sector. They come from different disciplines such as electrical, civil, mechanical engineering and should know about the features of hydro power plant covering the basics related to all the relevant disciplines and not just their own work. This book is targeted to present a good starting background for such professionals.

Chapter 1 of the book gives an introduction in the different types of power stations and overview of the status of hydro power worldwide. Chapter 2 presents basic terminology and legal issues related to the use of hydropower. In Chap. 3, the basic concepts of hydropower and the theory behind its utilization are explained. Chapter 4 covers an explanation of the major plant components. Since the turbine is the most important one, Chap. 5 is geared at covering various aspects related to different types of hydro turbines and their working principles. Chapter 6 covers working principles and construction of different type of oceanic power plants. The economics of hydropower plants is considered in Chap. 7 of this book before presenting the outlook for the future of hydropower in Chap. 8.

Attempts have been made to include in this book the technological advancements up to the beginning of the year 2011.

The authors thank Mrs. G. Schultz-Herzberger for checking the English language of the manuscript and Mrs. A. Osenbrueck for her help in adapting material from lectures of the authors to appear in this book.

Two very special thanks are given to Mrs. M. Koetter and to Mrs. M. Baerwinkel for their great help by typing, text formatting and preparation of figures and graphics.

Authors wish the readers of this book a happy plunge into the field of hydropower.

May 2011

Hermann-Josef Wagner
Jyotirmay Mathur

Contents

Chapter 1
Introduction and Status of Hydropower

1.1 Energy Forms

Nearly all the activities of human life nowadays, are dependent on some form of energy or another. Energy is the ability of a "system" to carry out work. Energy can be transferred from one system to another in any of three ways:

- by carrying out mechanical work, as in a belt drive
- by heat exchange, as in a steam engine
- by electromagnetic fields, as in an electromotor.

Physically speaking, energy can be converted from one form to another. Of particular practical importance are thermal and electrical energy.

The few preferred forms of energy are electricity, gas and oil. Energy can be available in different forms such as:

- potential energy: energy stored within a physical system as a result of the position (differences in altitude)
- kinetic energy: the energy possessed by a body because of its motion, equal to one half the mass of the body times the square of its speed
- thermal energy: the energy in any system by virtue of temperature
- electrical energy: the energy made available by the flow of electric charge through a conductor
- chemical energy: the energy due to associations of atoms in molecules and various other kinds of aggregates of matter
- nuclear energy: the energy released during a nuclear reaction as a result of fission or fusion. It is also called atomic energy

Today, a major portion of the world's energy requirements is met by energy sources which are burned and produce heat or thermal energy. Mankind is mainly dependent on the use of fossil fuels for this purpose; however, a small percentage of such energy is also obtained through biomass. The limitation with energy

H.-J. Wagner and J. Mathur, *Introduction to Hydro Energy Systems*,
Green Energy and Technology, DOI: 10.1007/978-3-642-20709-9_1,

released as heat is that it cannot be used without conversion for producing mechanical work such as running a pump or moving a car. Similarly it cannot be directly used for lighting an electric lamp or operating a computer. For this reason, heat is also termed 'low grade' energy.

On the other hand, in many cases, kinetic energy, usually in the form of rotation of a shaft, is a much preferred form of energy. According to the laws of thermodynamics, only a portion of heat energy can ever be converted into mechanical energy. This law is known as the second law of thermodynamics. As per a corollary of the second law, the portion of heat that can be converted into work is larger if the temperature at which heat is supplied is higher and the temperature at which the remaining heat is rejected to the atmosphere is lower. However, the maximum temperature is generally restricted by the ability of the materials to withstand the heat, whereas the lower temperature of heat sink is usually not lower than the temperature of the environment. Due to the above restriction, the remaining portion of heat energy, which is not converted into mechanical work, is dissipated into the environment as waste heat. These restrictions are the reason for the fact that the efficiency of most coal, oil or gas based power plants does not exceed 50%.

When compared to thermal energy; mechanical and electrical energy, also known as 'high grade energy', are preferred, since they can be converted into all other forms of energy with no major losses. Electricity is the most preferred since it can easily be transported over large distances through transmission lines, which becomes difficult for mechanical power.

In spite of all this, no energy form can be "produced" or "consumed" according to the law of energy conservation, but can only be converted from one form to another. The terms "energy generation" and "energy consumption" are commonly used in everyday life and the energy industry. Economically, what is involved is indeed the relationship between producers and consumers. Energy that has been "produced" is capable of doing several tasks, and the energy when considered to have been "consumed" means that it is in an economically worthless form. Therefore these terms will be used in this book in the same manner.

1.2 Energy Units

The energy industry and energy technology commonly use a large number of different energy units, which frequently makes it difficult compare data on energy consumption, energy requirements and types of energy sources used. For this reason, Tables 1.1 and 1.2 contain a list of frequently used units, prefixes and conversion factors.

Under the international unit system (SI) introduced in 1960, the Joule (J) and the kilowatt hour (kWh) derived from it are the mandatory legal units for energy.

The unit of power in the SI unit system is the Watt . The power of light bulbs is measured in Watts (W) that of cars in kilowatts (kW or 1,000 watts) and of power stations in megawatts (MW, or 1,000,000 watts).

Table 1.1 Conversion of energy units (see for text abbreviations)

Conversion Factors		
Unit	kJ	kWh
1 kilojoule (kJ)	–	0.000278
1 kilowatt hour (kWh)	3,600	–
1 MWannum	–	8,760,000

Table 1.2 Prefixes and abbreviations

Prefix	Abbreviation	Exponent	Number
Kilo	k	10^3	Thousand
Mega	M	10^6	Million
Giga	G	10^9	Billion
Tera	T	10^{12}	Trillion
Peta	P	10^{15}	–
Exa	E	10^{18}	–

The electric energy consumption in a home is indicated in kilowatt hours (kWh). For example, an electric device of 1,000 W operated for 1 h consumes 1 kWh. A typical household uses between 3,000 and 6,000 kWh per year in developed countries, whereas this number is relatively smaller for households of developing countries. On the other hand, the electric energy generation of a country is measured in terawatt hours (TWh)—1 billion kilowatt hours equal one terawatt hour, e.g. Germany had an electric energy generation of about 630 TWh in the year 2010.

1.3 Water Cycle in Nature

Water available in the universe goes round and round through different forms and phases through a process called the water cycle, as shown in Fig. 1.1. In some parts of the cycle, water is a liquid (rain). In other parts it is a gas (water vapour) or a solid (ice). The heat of the sun vaporizes water from seas, rivers and lakes and also from the soil and plants on the land. This water turns into an invisible gas called water vapour through a process called "evaporation". The water vapours become cooler as they rise into the atmosphere.

Since the moisture holding capacity of cool air is much less than that of warm air, upon rising high, some of the vapours turn into water droplets. This process is called "condensation". In the sky the tiny water droplets form clouds. When these droplets combine to form larger droplets, due to their weight, they fall to the earth as rain, hail or snow. Much of the water that falls on the land flows to the sea in streams and rivers. Some gets soaked into the ground and some stays as ice. The water eventually finds its way into rivers and seas, where the water cycle starts all over again.

During the water's journey to the sea, its energy is used for generating power through hydro power plants. In a way, hydro power plants can be termed as

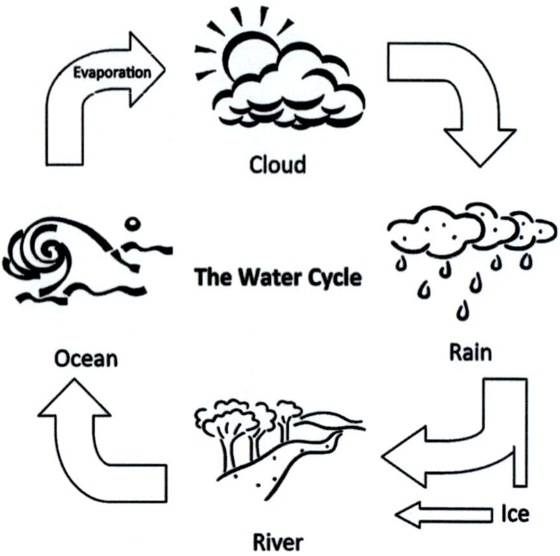

Fig. 1.1 The water cycle in nature

'man-made obstructions' in the path of water after having fallen as rain or snow and before flowing into the sea.

All the forms of energy on earth beside nuclear fuel, including hydro energy, are considered to be derived from solar energy. As can be seen in Fig. 1.2 out of 3.9×10^6 EJ of energy coming from the sun to the earth every year, about 22% of solar energy is consumed for the formation of rain which becomes the main source of hydro power.

1.4 Introduction to Hydropower

Man has been using water power since the beginning of civilization. Along with the burning of wood for light and heating, water power was used as the main source for generating mechanical driving power. The water streaming down from higher to lower levels consists of potential energy in itself because of its altitude which is converted into kinetic energy while flowing downhill. Jointly these energy forms contribute to what we call water power. It is a renewable source of energy because it is renewed continuously in a natural way.

1.4.1 Classification of Hydropower Plants

Hydropower plants can be categorised by different aspects. According to the working of hydropower plants, for example, they can be categorised by the source of

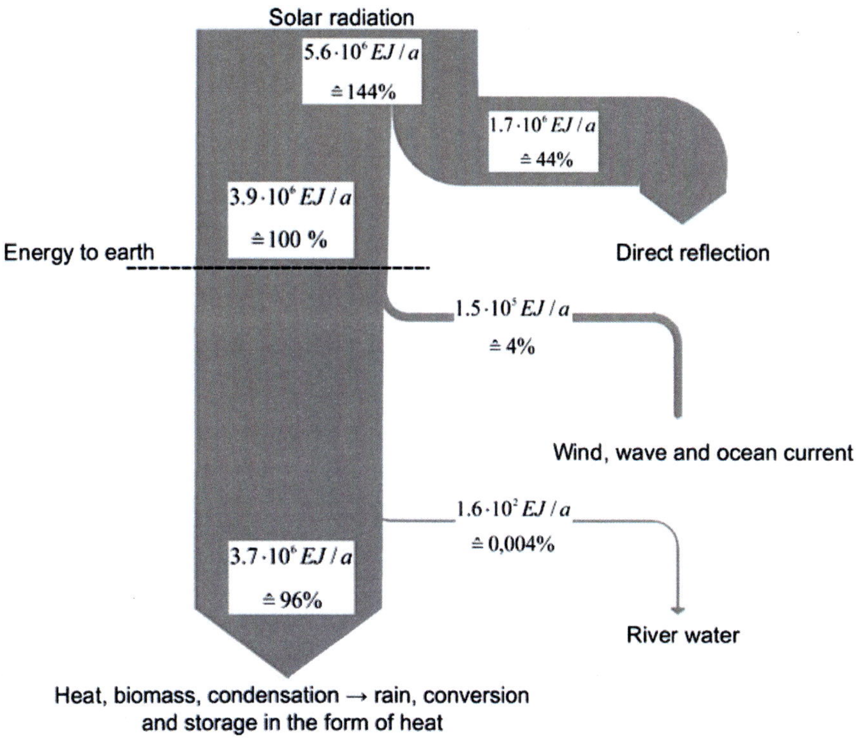

Fig. 1.2 Energy flow diagram of solar radiation

water, by their type of construction or also by their turbine as is shown in Fig. 1.3. This categorisation helps to better understand various aspects of a hydropower plant and to understand how some categories are connected to each other.

1.4.1.1 River Power Plants

River power plants are also termed run-of-river power plants in some countries. In this type of plants, the natural flow and elevation drop of a river are used to generate electricity (Fig. 1.4). Some of these power plants are fed directly by a river, whereas others are fed by a diversion canal as shown in Fig. 1.5. The latter power plants are usually small in power generation capacity. Diversion means that the power plant is not fed water by the whole river but that a part of the river water is separated in a canal. The canal feeds water to the little power plant as shown in the figure. Due to the limitation of flow in the diversion canal, a diversion or canal power plant cannot produce as much electricity as a power plant in the river itself.

The operating mode of such plants can vary in three different aspects. First, the river plant can work with a weir and without a slack flow. Here the river water streams directly through the turbine without being stopped by a dam.

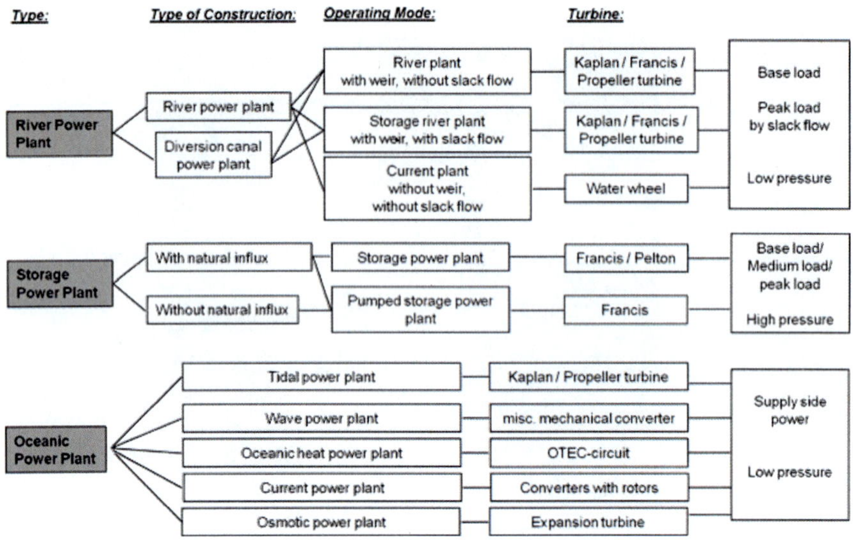

Fig. 1.3 Overview of hydropower plants

Another mode of operation of river power plants is the operation with little storage, which has a weir as well as a slack flow. The water is stored in a small reservoir. There are two reasons to store water. The first reason is to have a certain depth of water to make the river navigable for ships. The other reason is to have water reserves in order to serve as peak load. Capacity of such reservoirs is much smaller as compared to reservoirs of storage type hydro power plants.

The third mode found in small river power plants lacks both weir and slack flow. Looking at the category "turbine" in Fig. 1.3 one understands that a water wheel does not need weir or slack flow as it runs just by the streaming (surface) water of a river. Only the smaller power plants, like the water wheel, do not have a weir.

The turbines that are used for the first two operating modes are Kaplan turbines, Propeller turbines and Francis turbines. These are suitable for a relatively small height of fall and huge masses of water. That means that the pressure is relatively low. The pressure of a water pile of 10 m height has about 1 bar : One bar equals 0.1 Megapascal (MPa). A height of fall of 6–7 m means a difference in pressure of 0.6–0.7 bars (plus the atmospheric pressure, which is about 1 bar). The function of this type of hydroelectric power plant is mainly to deliver energy for the base load. Capacity of a river power plant can range from a few kilowatts to several hundred megawatts, depending upon the volume of water and height available for the fall of the water.

The advantage of river power plants over storage power plants is that since there is no or only a small reservoir, people living at or near the river do not need to be relocated and natural habitats are preserved, thus reducing the environmental impact. The disadvantage of this type is that the output of such plants is highly dependent on the river run-off which may not match with the power demand.

Fig. 1.4 River power plant in southern Germany in winter time (capacity about 4 MW) (photograph by courtesy of R. Lenk)

Fig. 1.5 Scheme of diversion canal fed river power plant

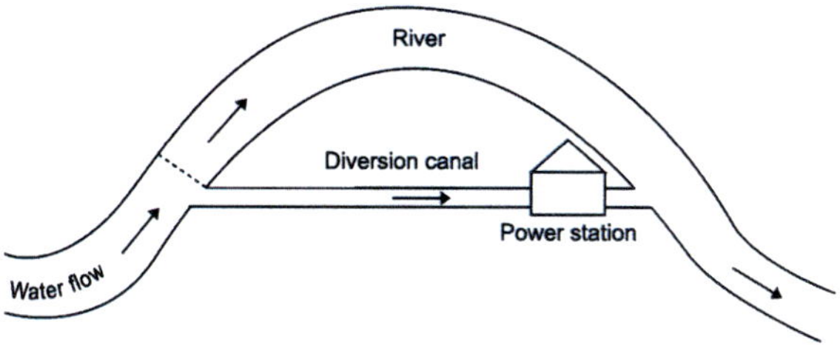

Sometimes, river power plants become the only choice in situations where e.g. a river flows from one country to another country and where by agreement, stoppage of flow by one country is not allowed for reasons of security, safety, irrigation or drinking water availability in the other country.

1.4.1.2 Storage Power Plants

Another major type of hydropower plants is the storage power plant (Fig. 1.6). This type of power plants may or may not have a natural influx of water. A natural influx can be a rain fed river or collection of melted water draining from mountains.

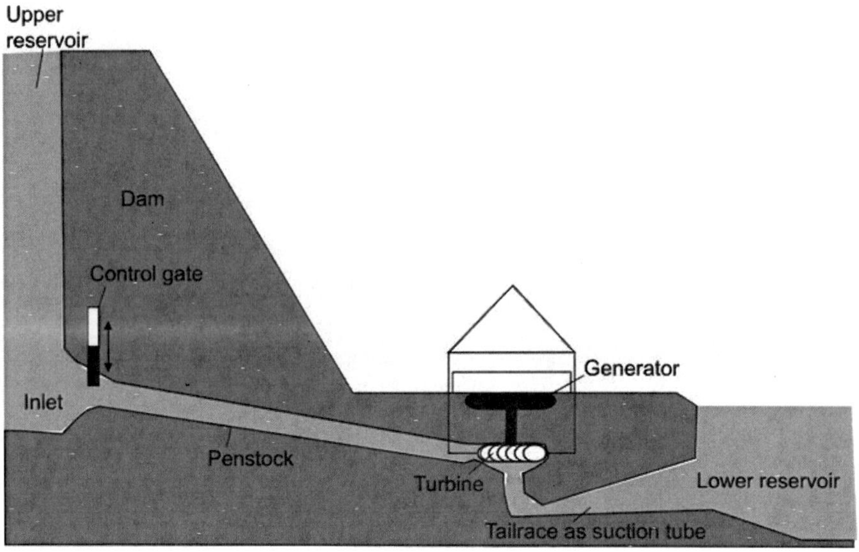

Fig. 1.6 Scheme of storage power plant

The bodies of water in the different spots between the mountains are connected through pipes. This is done to make the gaining of water, which is potential energy, as efficient as possible, e.g. in order to avoid draining of water, which would mean to lose energy.

If there is no or not enough natural influx, the power plant operates necessarily as a pumped storage power plant, the concept of which is explained later in this section.

The turbines that are used in storage type power plants are Francis turbines and even Pelton turbines. Francis turbines are used until fall heights of about 15–500 m, while Pelton turbines can work with fall heights of up to 2,000 m. On the other hand, Francis turbines can manage a huge amount of water flow, while Pelton turbines work with a relatively much smaller flow. Kaplan turbines are not commonly used in storage power plants, as they work well until heights of about 25 m. The concept of turbine suitability has been elaborated in subsequent chapters.

As one can imagine, the pressure of water is definitely high. If we imagine a height of water in the mountains of 800 m down to the turbine, the pressure is 80 bar. Large amount of water with such high pressure has potential of generating large amount of power that may not be needed all time. The function of storage power plants, therefore, is to store the potential energy of water and allow its usage when needed, like in medium and peak load hours, and also to cover seasonal fluctuations in water availability when feeding rivers. Even the base load can be delivered only as long as enough water is available in the reservoir. Storage power plants can deliver base load for a few months because they can store the winter's melt water until summer in the reservoir behind a dam. A regular pumped storage power plant, on the other hand, cannot deliver base load, as the capacity of reservoir is too small which would serve for only a few hours of base load. The pumped stored water is mainly to be used for peak load hours.

1.4.1.3 Working of Pumped Storage Power Plants

Storage power plants collect the water in a reservoir located at a high altitude, from the creeks and rivers in a relatively large catchment area around the reservoir. Often, for conservation reasons, not all the water available in these watercourses is brought to the reservoir. The reservoir is built for two different reasons. First, it serves to store the potential energy of the water, which can be taken from the basin and fed to a power station at a lower level at any time. That makes it possible, for example, to use the water from the spring thaw to generate power in the fall. Second, the reservoir has the task of meeting peak demands for electric power which may appear at short notice. Industrial plants and households do not need the same amount of electricity around the clock. Plants and equipment are turned off at night, while cooking stoves are needed around noontime and when people come home from work in the evening; at that time, too, washing machines and dryers are loaded and TV sets switched on. Hence, there is additional need for consumption of electricity at certain times of the day, so that the power stations must provide more electrical output then. Such peaks are extremely noticeable during breaks in major sporting events. As long as the fans are watching the game, the apartments are dark and most of the electrical equipment is turned off. Then, at the beginning of halftime, the lights are switched on in the living room, in the refrigerators as the doors open, and in the bathrooms as the toilets are used. This happens at the same time in up to millions of apartments, so that the demand for electric power shoots up in seconds. Pumped storage hydropower plants are there to provide the technical solution—additional power, fast. The slide in front of the turbine is opened, and within just 1–3 min, the turbine is turning and the generator starts producing power. Other possibilities for producing electric power quickly are gas turbines, which are similar to the turbines seen in airplanes, for example. They, too, can start fast and help meet such demand peaks.

Where there is little natural water available to draw on, a different kind of storage power plant, i.e. the pumped storage plant is used. Here, two reservoirs are built, one in the valley and one on the mountain, and the water in these reservoirs is used to generate electricity. If much electric power is needed, water from the upper basin is allowed to flow down to the lower one through the turbine after having generated power. If, on the other hand, little power is needed, for example at night, power is taken from the grid to pump the water up again from the lower reservoir to the higher reservoir to make more water available on the next day. Large pumped storage plants which operate in this way include the oldest pumped storage plant in the world at Herdecke on the Ruhr, Germany, which has been in operation since 1920, the large system at Vianden, Luxemburg and one of the most modern pumped storage plants worldwide in Goldisthal, Thuringia, Germany.

Pumped storage power plants are used for two purposes. The most common use is for meeting peak loads, whereas the second purpose of these plants is for providing reactive power. For physical reasons, the electricity consumers actually need not only electrical energy; they need it in two specific forms: first, the effective power or active power, such as power for producing the heat for a stove, or driving power of the motor. In order to turn the electric motor, magnetic fields must be built up and then turned off again for making the reactive power available. The generator must provide this reactive power too, at the quantity required by the consumer at all times. All this goes in a cyclic manner and the plants sometimes undergo even more than fifty cycles every day. This means, over the course of twenty-four hours, the power station will change the operation up to fifty times between pumping water, producing active power and producing only reactive power.

For this reason, in developing and underdeveloped countries, since there is deficiency of power, pumped storage systems are used mainly for meeting the peak demand of power, and the reactive power is provided through the use of electric circuits.

While a run-of-river power station and a storage power station without pumping is a real energy producer, a pumped storage plant which only serves to meet demand peaks, with no inflow into the upper reservoir, is actually an energy consumer. Power is used to pump the water up the mountain, possibly from coal-fired power stations. The pumps cannot operate without loss, nor are the electrical propulsion of the pumps or the pipes loss-free. If the water runs down the mountain again, it suffers a loss of energy due to friction in the pipes. After that, it drives the turbine, which is also not quite loss-free. The turbine which drives the generator, is not loss-free either. Of every 100 kWh of electrical output taken from the grid to power the pump, about 20 kWh and more will be converted to technically useless heat, according to the laws of physics, and ultimately, at a different time of day, the remaining 80 kWh can be fed by the generator into the grid as electric power (see Fig. 1.7).

The overall efficiency of a hydro power plant is therefore a product of individual efficiencies that can be expressed by the equation given below:

$$\eta_{overall} = \eta_{transformer} \cdot \eta_{generator} \cdot \eta_{turbine} \cdot \eta_{pumps} \cdot \eta_{others}$$

With today's state of the art, pumped storage plants are the only way to store large quantities of electrical energy using the detour of potential energy. No other storage medium can do so at such a scope. This is the reason why we are discussing pumped storage plants in the context of the use of fluctuating renewable energy sources like wind. On the one hand, it would be possible to pump the water into the upper basin with the help of a surplus of electric power from wind energy systems. That would permit a decoupling of the wind-power supply from the demand side. On the other, more pumped storage plants could be built to be used

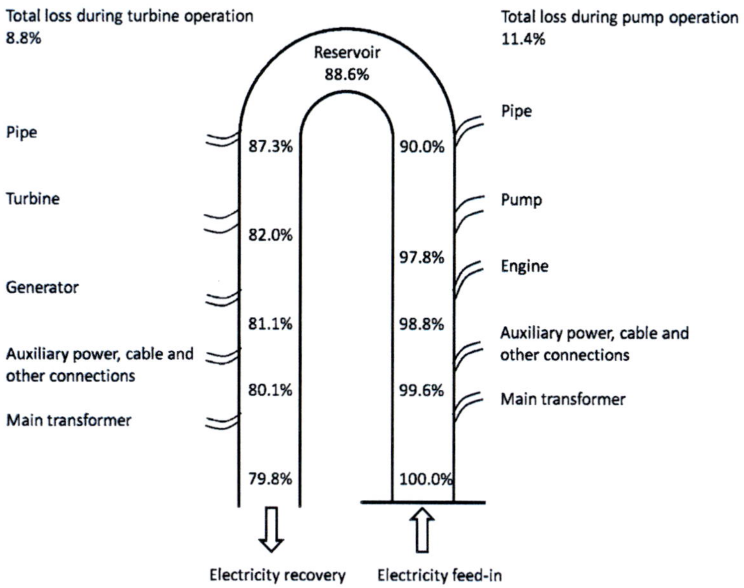

Fig. 1.7 Sample energy balance of a pumped storage power plant

if, due to major fluctuations of the wind, the output of conventional power stations had to be increased or reduced very rapidly. However, it must be taken into account that good wind sites are often far away from the low mountain ranges in which pumped storage plants can best be built.

1.4.1.4 Oceanic Power Plants

The third type of hydropower plants shown in Fig. 1.3, are oceanic power plants. This category of power plants shows a great variety of types of construction.

The tidal power plant, to begin with, is nearly like a river power plant with a dam. However, here the water from the ocean side can be stored behind a dam during the high tide and is stored from the other side during low tide. The turbines which are set in this power plant are Kaplan or Propeller turbines.

Another type of construction is the wave power plant. Wave power plants convert the potential power of waves mechanically through some mechanisms as shown by the example in Fig. 1.8. It converts the wave energy in pressure of fluid using a pendulum door (left scheme). Another principle is that wave water pressed air which is going through a wind propeller (right scheme).

The oceanic heat power plant, another type, works according to the OTEC principle. This is an abbreviation of "Ocean Thermal Energy Conversion" and uses the difference in temperature between surface water and deep water, and runs a circuit to gain electricity. Today, however, there is no OTEC power plant in commercial operation.

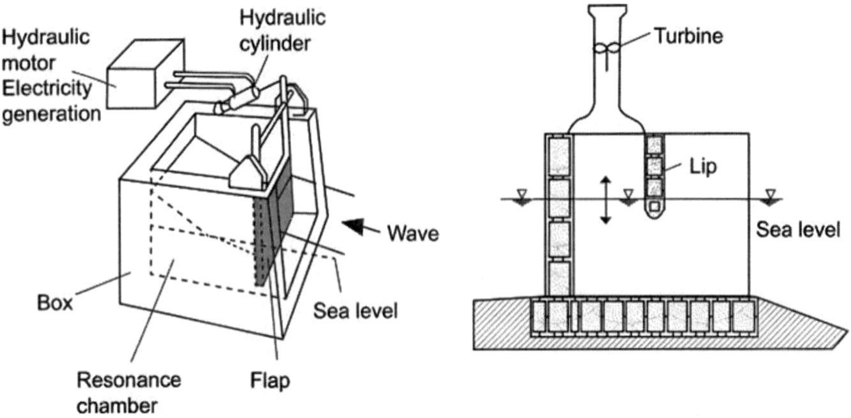

Fig. 1.8 Schemes of wave power plants (*left* Yagishiri/Japan, *right* Trivandrum/India)

A fourth type of oceanic power plants is the current power plant. This type works like a wind energy plant but under water. The energy of the streaming water is simply converted by rotors like a wind converter offshore.

The current power plant uses the movement of water in oceans originating from differences of water densities due to temperature differences.

A further type of oceanic power plant is the osmotic power plant. Here energy is gained from the difference in salinity of ocean water and river water which produces pressure in the water. The pressure is converted to rotating energy by an expansion engine drives the generator.

Generally, except for the electricity generation by tidal turbines—the generation of electricity in ocean power plants is still in a stage of technical development.

1.4.2 Classification Based Upon Power Generation Capacity

Facilities range in power generation capacity from large power plants that supply millions of consumers with electricity to small and micro plants that individuals operate for their own energy needs or to sell power to utility companies. The three categories of hydro power plants according to their power generation capacity are as follows:

1. Large hydropower
2. Small hydropower
3. Micro hydropower

Although definitions of these categories vary from one country to another, we suggest that large hydropower are facilities that have a capacity of more than 15–20 MW (up to several GW), small hydropower plants have capacities in the range of 0.1 to 15–20 MW, and micro hydropower plants have capacities of less than 100 kW.

Table 1.3 Global use of hydropower in 2006 (Numbers are rounded. *Source* UN, quoted recording VIK statistics [1])

Country	Hydro Power Generation (TWh)	Share of total generation (%)
Austria	40	59
Brazil	350	83
Canada	360	58
China	440	15
France	60	11
Germany	30	4
India	110	15
Japan	100	9
Norway	120	99
Russia	180	18
USA	320	7
Other	1010	
World	3120	17

However, there is no worldwide consensus on the definition of small hydropower: Some European countries such as Portugal, Spain and Ireland accept 10 MW as the upper limit for installed small capacity. In Italy power stations with more than 3 MW are expected to sell their electricity at lower prices and in Sweden the limit is 1.5 MW. In France, the limit has recently been established at 12 MW, not as an explicit limit of small hydropower stations, but as the maximum value of installed power for which the grid has the obligation to buy electricity from renewable energy sources. In Great Britain, 20 MW is generally accepted as the threshold for small hydro. In Germany, the capacity of hydropower is important for the feed in tariff. It is different for hydropower stations below 500 kW, from 501 kW to 2 MW, and over 2 MW. In India, 15 MW is the limit for small hydropower plants.

1.5 Status of Hydropower Worldwide

Some 16% of the electric energy produced worldwide is from hydroelectric facilities. In some countries, it is the most important source of electricity. Norway gets 99% of its electric power from water, Brazil 84% and Canada 58% as shown in Table 1.3.

On the basis of the installed power generation capacities, the world's biggest storage power stations are the Three Gorges Dam in China, the Itaipu Dam in South-America, on the border between Brazil and Paraguay, and several other dams as shown in Table 1.4. These huge dams provide electrical outputs up to twenty-five times greater than those of a single unit in a coal-fired power station. The output from Itaipu would statistically suffice to supply Paraguay and over 20%

Table 1.4 World's largest hydro power stations

Power plant	Installed capacity (MW)
Three Gorges Dam (China)	22,500
Itaipu Dam (Brazil/Paraguay)	14,000
Xiluodu Dam (China)	12,600
Belo Monte Dam (Brazil)	11,000
Guri Dam—Simón Bolívar hydroelectric power station (Venezuela)	10,240
Tucurui Dam (Brazil)	8,550
Grand Coulee Dam (USA)	6,810
Sajano-Schuschenskaja GES (Russia)	6,400
Longtan Dam (China)	6,300
Xiangjiaba Dam (China)	6,000
Krasnoyarsk Hydroelectric Dam (Russia)	6,000
Nuozhadu Dam (China)	5,850
Robert-Bourassa hydroelectric power station (Canada)	5,620
Churchill Falls hydroelectric power station (Canada)	5,430
Jingping II Hydropower Station (China)	4,800

Table 1.5 World's largest pumped hydro power stations

Power plant	Installed capacity (Turbine power) (MW)
Kannagawa Pumped Storage (Japan)	2,820[a]
Bath County (USA)	2,730
Robert Moses Niagara Hydroelectric Power Station (USA)	2,520
Guangzhou Pumped Storage Power Station I + II (China)	2,400
Huizhou Hydroelectric Power Station (China)	2,400
Dneister Pumped Storage Plant (Ukraine)	2,270[b]
Okutataragi Pumped Storage Power Station (Japan)	1,940
Ludington Pumped Storage Power Plant (USA)	1,870
Tianhuangping Pumped Storage Power Plant (China)	1,840
Grand Maison (France)	1,800
Dinorwig Power Station (Wales)	1,730
Raccoon Mountain Pumped Storage Plant (USA)	1,600
Kazunogawa Hydroelectric Power Plant (Japan)	1,600
Mingtan Power Plant (Taiwan)	1,600
Tumut 3 (Australia)	1,500

[a] Completion 2016
[b] Completion not yet reached

of Brazil with electrical power. Hydroelectric power systems of this order of magnitude have considerable impacts on nature and also on the living space of people. According to press releases, one million people have been resettled for the Three Gorges Dam project in China. Such large quantities of stored water are also a potential threat, if the dam wall breaks; vast areas would be inundated and many people endangered. Renewable energies used locally on such a scale are no longer environmentally neutral. Even run-of-river power stations require a number of ecological compensation measures. These include near-natural fish stairways and spawning waters, or the creation of flood-plain forests.

Beside the storage power stations there are also many big pumped storage power stations in operation worldwide. Table 1.5 shows the most important ones.

Features of Hydropower Plant at Three Gorges Dam, China

The Three Gorges Dam spans the Yangtze River by the town of Sandouping, located in the Yilling District of Yichang, in Hubei province, China. It is world's largest electricity generating plant. Since 2009 the project has produced electricity, it increases the river's shipping capacity, and reduces the potential for floods downstream by providing flood storage space.

(photograph by courtesy of G. Subklew)

Data of Three Gorges Dam Hydropower Station

Year of beginning of construction	1994
Installed power	22,500 MW
Number of turbines	34
Type of turbines	Francis
Total water reservoir capacity	40 km^3
Rated power per unit	700 MW (×32), 50 MW (×2)
Length of dam	2,300 m
Maximum height of dam	185 m
Produced electricity (2009)	≈80 TWh

Features of Hydro Power Plant at ITAIPU, South-America

The Itaipu hydropower plant is a joint venture of Brazil and Paraguay. It uses the water resources of the Paraná River.

In Terms of yearly produced hydroelectric energy, the Itaipu power station is the largest hydropower station in the world.

The generated electricity is split up. On half is delivered to Brazil (60 Hz frequency) supplying the region of Rio Grande do Sul, Paranà, Sao Paulo and Rio de Janeiro, the other half is delivered to Paraguay (50 Hz frequency), which needs only small amount of this. The rest is converted from 50–60 Hz frequency and sold to Brazil. Itaipu electricity covers about 20% of electricity demand of Brazil and about 90% of the demand of Paraguay.

Data of ITAIPU Hydropower Station

Year of beginning of construction	1975
Installed power	14000 MW
Number of turbines	20
Type of turbines	Francis
Rated water flow per unit	645 m^3/s
Rated power per unit	715 MW
Number of poles (50 Hz/60 Hz)	66/78
Length of main dam	610 m
Maximum height of main dam	196 m
Produced electricity (2007)	≈90 TWh

The "largest" European hydroelectric power stations are, by worldwide standards, small. An electricity generating unit in a modern hard-coal-fired power station has an output of 700 MW; theoretically, it should be able to meet the requirements of a city of about 700,000 inhabitants in an industrialized country. The storage power stations Malta and Kaprun in Austria are in this output range. The largest run-of-river hydropower station in Germany however, has an output of only about 20% of this level. This is simply due to the available water supply and the usable height difference. Rivers like the upper Rhine are therefore covered with a cascade of power stations, one after the other. On the Moselle and other rivers, there are ship sluices, so that the ships can overcome the difference in altitude of the dammed water; some rivers were made navigable in the first place by these means.

The biggest pumped hydro power stations in Europe are Vianden in Luxemburg and Goldisthal in Germany, each generating a little bit more than 1,000 MW power.

In Germany, hydroelectric facilities provide about 4% of electricity consumption. Moreover, there is little scope for further development of this source in Germany; on the other hand, worldwide hydroelectric power capacities are far from being exhausted yet. This does not necessarily mean the construction of new hydroelectric power systems, but largely the modernisation and expansion of existing systems. Rebuilding the hydroelectric power station at Rheinfelden on the upper Rhine River for example will be able to quadruple output from 26 MW to 100 MW. During the 1990s, many small hydroelectric power stations which had previously been shut down were reactivated; hence, almost 5,000 small hydroelectric power systems—"small" meaning less than 1 MW—are in operation in Germany. However, by far the major share of the 4% of Germany's electricity provided by hydroelectric power does not come from these privately operated units, but from about 120 larger systems—i.e., those producing more than 382 5 MW—located on the larger rivers and operated by the power companies. Due to their size and the greater availability of water, they provide 80% of the annual production of hydroelectric power (see Table 1.6).

The situation and development of the hydropower sector in developing countries are little different from developed countries. The main difference is that there are different types of technological, economical and social issues that are to be addressed. For example, India is blessed with immense amount of hydro-electric potential and ranks 5th in terms of exploitable hydro-potential on a global scenario. It has an economically exploitable and viable hydro-potential assessed to be about 84,000 MW at 60% plant load factor (149,000 MW installed capacity). In addition, 56 sites for pumped storage schemes with an aggregate potential of 94,000 MW have been identified. Further, hydro-potential from small and micro schemes has been estimated at about 6,800 MW from 1,500 sites. Thus, in totality India is endowed with hydro-potential of about 250,000 MW. Major potential of hydro power lies with basins of main rivers as given in Table 1.7. However, exploitation of hydro-potential has not been up to the desired level due to various constraints confronting the sector. Only about 20% of the potential has been harnessed till 2010. The constraints which have affected hydro development are technical (difficult investigation, inadequacies in tunnelling methods), financial (deficiencies in providing long term financing), tariff related issues and managerial weaknesses (poor contract management). The hydro projects are also affected by geological surprises (especially in the Himalayan region where underground

Table 1.6 Hydropower plants in Germany in 2005

	Number of plants	Net capacity (MW)	Net generation (TWh/annum)
Under 1 MW	6,900	530	2.3
Over 1 MW	400	3,400	17.3
Total	7,300	3,930	19.6

Table 1.7 Hydropower potential in India (*Source* NHPC, India [2]) without pumping hydropower stations and small hydropower stations

Basin/Rivers	Probable installed capacity (MW)
Indus basin	34,000
Ganga basin	21,000
Central Indian river system	4,000
Western flowing rivers of southern India	9,000
Eastern flowing rivers of southern India	15,000
Brahmaputra basin	66,000
Total	149,000

tunnelling is required), inaccessibility of the area, problems due to delay in land acquisition, and resettlement of project affected families.

In 1998, the Government of India announced a "Policy on Hydro Power Development" under which impetus is given to the development of hydropower in the country. This was a welcome step towards an effective utilization of water resources in the direction of hydropower development. During October 2001, the Central Electricity Authority (CEA) came out with a ranking study which prioritized and ranked the future executable projects. As per the study, about 400 hydro schemes with an aggregate installed capacity of 107,000 MW were ranked in A, B & C categories depending upon their inter-se attractiveness. During May 2003, Government of India launched the 50,000 MW hydro initiative in which preparation of Pre-Feasibility Reports of 162 Projects totalling to 50,000 MW was taken up by CEA through various agencies. (source: NHPC, India) [2].

In addition, the Clean Development Mechanism (CDM) has also helped the growth of the small hydro sector in India through improving their financial viability.

1.6 Advantages and Disadvantages of Hydropower

The advantages of hydropower can be summarized as follows:

- It is a renewable source of energy—and saves scarce fuel reserves.
- It is a clean power source, because there is no air pollution or radioactive waste problems associated with it.
- Since water power produces no carbon dioxide, it does not contribute to global warming.

- Hydropower stations have an inherent ability for instantaneous starting, stopping, load variations etc. and help in improving the reliability of power systems. As a result, hydro stations are the best choice for meeting the peak demand.
- Hydroelectric projects have a long useful life extending over 50 years. Some hydro projects completed at the end of the 19th century are still in operation (e.g. the plant installed in 1897 in Darjeeling, India).
- Average cost of generation, operation and maintenance over lifetime is lower than any other sources of energy.
- Hydropower has a higher efficiency (over 90%) compared to thermal energy (up to 45%) and gas (up to 60%).
- Cost of generation is free from inflationary effects after the initial installation.
- Storage based hydro schemes often provide attendant benefits of irrigation, flood control, drinking water supply, navigation, recreation, tourism, pisciculture etc.
- The location in remote regions leads to the development of backward areas inland (education, medical services, road communication, telecommunication etc.). This advantage is very important in developing and underdeveloped countries.

In spite of these advantages, there are a few limitations or disadvantages of hydropower, especially of large hydro power plants:

- The availability of hydro power is restricted to hilly or foothill areas due to the availability of water and head. This requires extra investment for installing long transmission lines and inevitably leads to transmission losses.
- Large dams are considered to cause a heavy concentrated load on the earth leading to seismic effects.
- Rehabilitation and restoration of people and activities in submerged areas is always a matter of concern.
- In underdeveloped countries the construction period of hydropower plants is longer than that of other plants. This is especially due to the fact that in hilly areas it is difficult to transport heavy equipment and machinery required to build the plant.
- During the rainy season, heavy rains in the catchment areas could become a risk to the safety of the dam. At the same time, the release of large amounts of water is also not possible since it creates floods in the downstream side.
- Some hydropower plants, especially in developing and underdeveloped countries, face serious problems due to an uncontrolled development in the catchment areas. This happens due to improved living conditions and employment opportunities in the area surrounding the power plant and the dam. The construction of housing and other commercial activities in the catchment area, over time start causing disturbances to the flow of water into the reservoir, which changes the availability of water for power generation.
- At some locations silt in the water due to soil erosion causes damage to the turbine blades requiring frequent maintenance during the rainy season.

References

1. VIK-Statistik der Energiewirtschaft (Statistics of energy economy) (Ed) VIK Verband der industriellen Energie–und Kraftwirtschaft e.v. (http://www.vik.de), Essen (Germany), June 2010 ISSN 0585-2005
2. a) National policy for hydropower development, Report from the national hydropower corporation, Government of India. b) Policy on hydropower development, Policy document from Ministry of Power, Government of India, 1998. c) Hydro development plan for 12th five year plan, Central electricity authority, Government of India, New Delhi, 2008 http://www.nhpcindia.com/writereaddata/English/PDF/hydro-policy.pdf

Chapter 2
Terminology and Legal Framework

2.1 Important Parts of a Hydropower Station

It would exceed the scope of the present work to mention the thousands of parts and components of a hydropower station, but the major parts are listed in Table 2.1. Since the presence of any part depends upon the type of plant, a detailed discussion on the function of these parts is presented later, in Chaps. 4, 5 of this book.

2.2 Operational Terminology

For the purpose of improving the understanding of the construction and functioning of hydropower plants, a brief description of commonly used terms related to power generation is necessary. An explanation of commonly used terms in hydro power is given below.

The following terms are used quite frequently in relation to power generation: generation:

Construction capacity. This is the amount of power generation capacity that can be achieved in a river power plant. This capacity depends on the construction flow rate at the construction drop height.

Maximum capacity. Concerning river power plants this is the same capacity as the construction capacity. Storage as well as pumped storage power plants have the maximum capacity at the highest adjustable capacity at maximal fall height.

Rated power. This is the highest continuous power of the machine which was agreed in the contract between operator and recipient. It is also known as power rating of any machine or plant, usually at this loading, machine performs the best.

Peak load supply. When there is a high demand of electric power, the hydroelectric power plant, mostly storage power plant, is used to cover the gaps between

H.-J. Wagner and J. Mathur, *Introduction to Hydro Energy Systems*, Green Energy and Technology, DOI: 10.1007/978-3-642-20709-9_2,

Table 2.1 Important parts of different hydropower plants

Name of part	Role	River plant	Storage plant (without pump)	Pumped storage plant	Oceanic plant
Turbine	Conversion of power of water into mechanical energy by rotation of shaft	Yes	Yes	Yes	Yes (only tidal)
Generator	Conversion of the mechanical energy of rotation of shaft into electricity	Yes	Yes	Yes	Yes
Spiral casing (also trumpet inlet)	Impinge water on the turbine symmetrically	Yes	Yes, if Francis turbine	Yes, if Francis turbine	Yes (only tidal)
Dam	Block flow of water for use in season of less water flow or high demand		Yes	Yes	Yes (only tidal)
Spillway	Allow excess water to flow downstream without producing power	Yes	Yes	Yes	Yes (only tidal)
Stilling basin	Reduce velocity of fast flowing water		Yes	Yes	
Penstock (see pressure pipe)	Delivers water from reservoir to turbine	Yes	Yes	Yes	Yes (only tidal)
Guide blades (guide vanes)(adjustable blades)	Ensures entry of water in optimal direction into moving turbine and open and closed water flow	Yes	Yes	Yes	Yes (only tidal)
Surge chamber	Buffers water by closing and opening of control valve		Yes	Yes	
Control gate (sluice)	Open and closed flow of water into penstock	Yes	Yes	Yes	Yes (only tidal))
Control valve	Open and closed water flow from outlet of penstock and controls the flow rate to inlet of turbine		Yes	Yes	
Tailrace	Carries outlet water from turbine to downstream main river	Yes	Yes	Yes	
Upper reservoir	Store water for use in season of less water flow or high demand	Yes	Yes	Yes	
Lower reservoir	Store water going out of turbine for reuse			Yes	

(continued)

Table 2.1 (continued)

Name of part	Role	River plant	Storage plant (without pump)	Pumped storage plant	Oceanic plant
Pump	Pump water from lower reservoir to upper reservoir for power generation during peak demand			Yes	
Pressure pipe	Carry water during pumping from lower reservoir to upper reservoir or carry water from upper reservoir to turbine (penstock)		Yes	Yes	
Hydraulic mechanism	Conversion of oscillation to rotation				Yes (only waves)
Cavern	Allow power plant to be situated inside mountain		Yes (possible)	Yes (possible)	

the demand and the supply of conventional power plants. This is done by running the machine up and down. The advantage is that the hydropower plant can be run up and down very quickly. Thus it can be used when there are sudden and quick changes in the load of the grid.

Height of fall, fall height, head. When using Kaplan, Francis and Propeller turbines, this is the difference in elevation between the water level of the upstream side (in the case of river power stations or on the upper reservoir in the case of storage power stations) and the elevation of the water on the downstream side (river elevation or lower storage elevations).

When using Pelton turbines this is the difference in elevation between the water level on the upstream side and the elevation of the water inlet to the turbine (nozzle).

Phase shift operation. The consumer needs in addition to active power also reactive power for driving motors or operating electronic devices with capacitance. Also long lines of cables or overhead lines need reactive power during operations. For delivering only reactive power, the generator is connected with the grid. By being over exited or under exited the generator delivers reactive power to the grid. The mechanical losses of turning generator and water empty turbine are covered by less active power which will is taken from the grid, also from other power stations. The turbine is not driven by water during the phase shift operation.

Regular working. Regular working means that the storage or pumped storage power plants are run at the load factor of the capacity-frequency-regulator. The machines of pumped storage plants permanently operate at different capacity, if there is need of peak load supply.

The change in capacity of river power stations is less than that of thermal power plants.

Regular year. A regular year is, with respect to available water, a statistical average year which is determined over a longer period, e.g. 20–30 years.

Reserve capacity. This is a type of supply which is quickly available and is used for changes of the load or for disturbances in other power plants or other power grids.

Firm capacity, firm power, secure power. Concerning river power plants, the firm power (sometimes also called secure power) is the operational capacity which is definitely available for a certain amount of days, e.g. in Germany 330 days. The number 330 is chosen arbitrarily. In power plants that have a short-time reservoir, e.g. a river power plant, the firm power is regulated by the capability to dislocate capacity. The available time of the firm power is also agreed between operator and recipient, e.g. 5 h daily.

Storage power plants have a different definition of firm power. Here it is the power which the power plant can deliver at fall height by a reservoir content of 10% of water volume. With pumped storage power plants the firm power is the delivered power by a fall height of reservoir content of 50% of water volume.

Turbine operation, generator operation. By turbine operation also called generator operation, the turbine runs the electric machine. The machine works as a generator and delivers the electric current to the grid.

2.3 Load Areas

During various hours of the day, the demand of electricity keeps varying due to a time based variation in the use of electric equipment in houses/offices and machinery in industries. In the morning of a working day the demand rises from about 6 to 8 a.m., when the majority of people get ready for work, causing a relatively high load on the power plants. Then the demand, and accordingly the power plant load, is relatively constant until noon (lunch time). Another peak load arises late in the afternoon or early in the evening when people come home and switch on their appliances and equipment such as lights, computers or TVs. Until night the load shrinks again. The highest load during the day is usually reached at noon or in the evening, while the lowest load occurs at night (between 2 and 4 a.m.) as shown in Fig. 2.1.

It must be noted that the figure only explains the variations a hydro power plant has to address. The demand and the variations significantly change with differences in location and climatic conditions. For example, in a warm country like India, the demand would be higher in summer due to requirements of cooling, while in Germany the demand is higher in winter due to heating requirements as already mentioned before.

Thus, the load not only varies throughout the day but also from season to season. In cold countries like Germany, the demand on electricity is high in winter, due to the need for more light and electricity for the heating equipment, for instance, but in summer less electricity is needed and, therefore, the constant load is lower than in winter. In warm countries like India, the trend is opposite; demand in summer is significantly higher as compared to winter due to an increased demand for air-conditioning.

Despite the variation in load over time, there a minimum load/demand that is always exists. This is termed base load. Base load or constant load is the level of load or demand on electricity that is definitely needed at every moment of the day and does not fall below that level. Middle load, on the other hand, means a load level that is demanded by the electricity consumers for several hours a day, usually from morning till late evening. In contrast to the base load (constant) the middle load is not defined by a constant level but can vary with the rise of electricity demand in the morning till the fall in the evening. The peak load, finally, is "the top of the middle load" that can graphically be separated from the average load and is reached at particular moments of the day.

Usually, a hydropower plant can have four different responsibilities related to power supply. The first responsibility is to produce power for meeting the base load, as explained above. This is mainly carried out by river water power plants. Task number two is to produce the peak load, which is fulfilled predominantly by storage power plants. The third task of hydropower plants in any country is to deliver reactive power whenever needed due to inductive load (see also Sect. 5.3.3), which is fulfilled by those pumped storage power plants that are especially equipped for

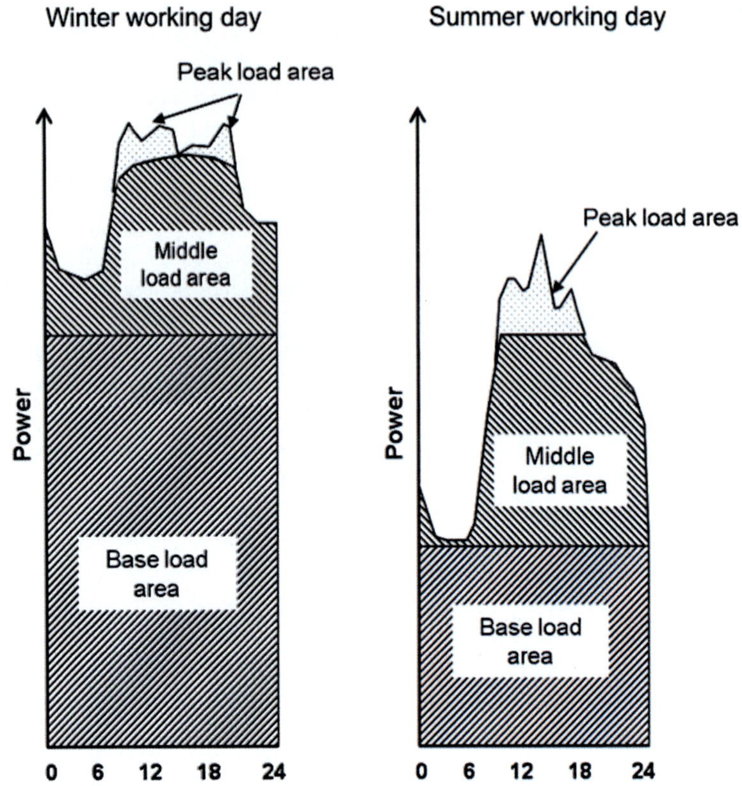

Fig. 2.1 Base, middle and peak load on a winter and a summer day of cold countries

this purpose. The last task, finally, is to store water for meeting seasonal variations in power demand and to overcome seasonal shortages in the availability of water.

2.4 Active and Reactive Power

If a hydropower station works in the operation mode of phase shifting, the generator delivers more or less reactive power to the grid.

Therefore, the concept of active power and reactive power will be explained below.

The time based variation of current $I(t)$ and voltage $U(t)$ in alternating current is sinusoidal (Fig. 2.2). Their relationships with time are given below:

$$I(t) = I_0 \cdot \sin(\omega \cdot t - \varphi) \tag{2.1}$$

and

$$U(t) = U_0 \cdot \sin(\omega \cdot t) \tag{2.2}$$

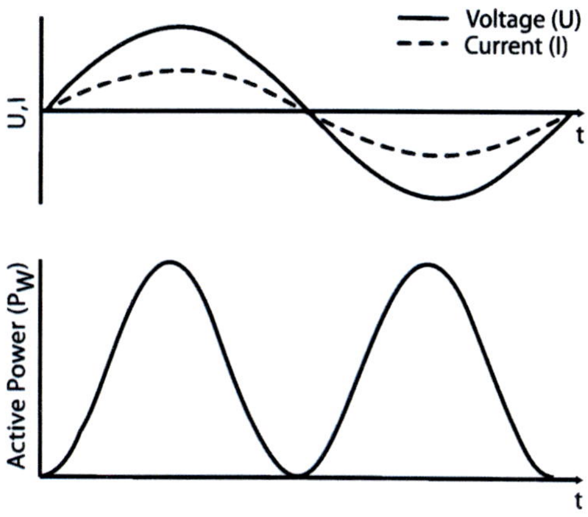

Fig. 2.2 Active power with voltage and current overlapping

In the above equations:

U_0 peak voltage,
I_0 peak current,
ω angular frequency (2πf),
φ is the angular phase shift of the electric current from voltage. In case of active power φ is, as mentioned above, zero

The active power P_W is given by:

$$P_W = \frac{1}{2} \cdot U_0 \cdot I_0 \cdot \cos\varphi \tag{2.3}$$

Electric and magnetic energy storage, like a condenser with the capacity C or a coil with the inductance L, use active power only marginally. If these are driven with the sinusoidal alternating voltage with the amplitude U_0, the current with amplitude I_0 and the angular frequency ω, an alternating power is flowing. Due to the sinusoidal variation of power, the condenser or the inductance is loaded or reloaded periodically.

The alternating power of this type is described as reactive power (see Fig. 2.3). In reality, there is some energy loss due to the loading and unloading cycles that result in transportation of energy and hence additional ohmic resistance losses in the grid. That is why reactive loads should be avoided as far as possible, or, expressed differently, reactive power should be compensated. At sinusoidal alternating current and AC voltage the reactive power P_R is given by:

$$P_R = \frac{1}{2} \cdot U_0 \cdot I_0 \cdot \sin\varphi \tag{2.4}$$

with the angular phase shift φ of the electric power opposed to the voltage.

Any end use of electricity requires an electromagnetic field a drill machine for instance, needs both active and reactive power. Active power is that part of the

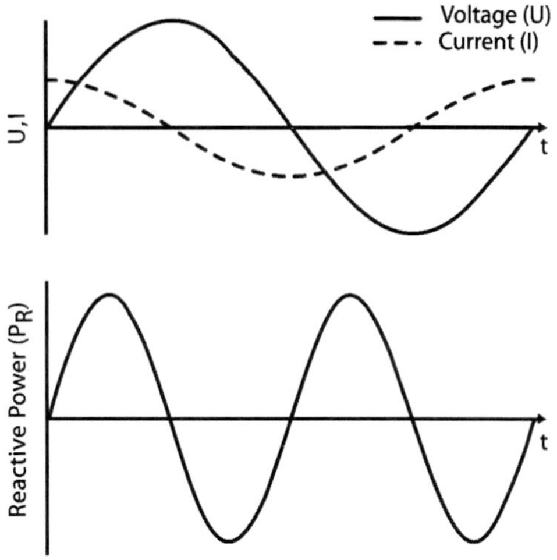

Fig. 2.3 Reactive power with voltage and current not overlapping

total power which is utilized as mechanical power. At active capacity electric power and voltage run in the same phase, meaning that the sinus curves overlap.

The reactive power, on the other hand, is used up by devices that require an electromagnetic field for their operation. Reactive energy is the energy in the magnetic field which rotates in the motor. The amount of reactive power required depends upon the strength of the magnetic field the motor needs. The shaft of a motor can only turn as long as an electromagnetic field is turning by setting up and down, therefore, this energy sways between the power plant and the electrical consumer having reactive power demand. The electrical consumer uses the reactive energy briefly and sends it back to the power plant over the grid.

The demand of reactive power could locally also be covered by switching a condenser together with an electromagnetic coil. This is due to the fact that an electromagnetic coil is setting up the magnetic field, when the condenser is setting down the electric field and vice versa. Energy could swap between the two by appropriate sizing of capacitance and inductance.

The grid operator must take care that the angular phase shift φ is not too big, in order to ensure the grid's stability. Therefore the grid operator needs power stations like pumped water power which are able to deliver reactive power.

2.5 Legal Requirements

For building a hydropower plant, in addition to the physical and economical considerations, legal considerations are equally important. In view of this different laws are described in the following sections:

2.5.1 Permission to Deviate Water

Water is essential for nature and human life. For centuries, it has been crucial for farmers to have rivers, brooks or even lakes on their property. However, this has often been a reason for conflicts. Farmers who do not have running water try to divert river water in nearby areas. Today, water related regulations in most countries do not only include restrictions on the diversion of running water, they also cover a variety of further regulations, like the usage of water, waterway construction, protection of water, handling of hazardous material etc.

For example, in some countries a landowner, who has a water stream/brook running through or along his property, is not allowed to use this water freely. He is required to obtain permission before creating any diversions or other modifications of the running water. This applies equally to local communities, states and the country as a whole.

Another legal requirement obliges the owner to keep the water clean. This is irrespective of having a hydropower plant on the water stream. If there is a hydropower plant on the stream, the owner of the power plant is not allowed to remove the waste in front of the dam wall and put it behind the wall.

2.5.2 Environmental Clearances

Any hydropower plant operator is required to consider the environmental impact of the plant in such a way that compensatory measures for the intervention upon the local nature by the construction of the power plant are taken into account.

In several countries, legal issues have constituted a major obstacle in the expansion of hydropower capacity. However, most countries like India and Germany have taken the following legal initiatives through framing a 'Hydro Policy' or 'Federal Water Act' or similar provisions.

In most countries the law provides for a test procedure to assess the impact on the environment that a construction like a hydroelectric power plant would have. If the impact on the environment is too severe, the impact has to be compensated by positive measures for the environment. The aim of such legal provisions is to guarantee uniformity in the principles for determining, describing, and judging the impact on the environment of certain plans that concern waters. The result of the environmental impact assessment has to be respected by all official decisions concerning the legitimacy. Upon request, the responsible authority assesses on the basis of the planned construction, if an environmental impact study has to be carried out. This assessment is to be made public.

The documents a planner submits to the authorities have to contain, e.g. a description of his plan regarding location, type, and extent of the project as well as the requirements on ground and soil. Also he has to describe measures that minimise the negative impact on the environment as well as the potential compensatory measures. Further, the planner needs to hand in a description of the environment and the environment's parts in the affected area of the plans. In addition, he has to present an overview of the other most important possible solution as well as a statement concerning the crucial reasons for selecting these proposals. Environmental impacts that affect neighbouring countries have to be considered in this procedure as well.

The general public is in so far involved in this process as the intention of the construction is announced publicly, the required documents are open for inspection, the public has the possibility to make comments, and finally, when the decision of the authorities is made public.

In India, similarly to many other countries, it is mandatory to receive environmental clearance from the Ministry of Environment and Forests, before construction of a hydropower station of the category "River Valley, Multipurpose, Irrigation and Hydro-electric Projects". In the Environment Investigation Analysis, all the alternatives explored by the project proponent are to be studied from the environmental angle. Scoping matrix, likely impacts identified for various environmental aspects (geological, biological, seismic, hydrological, fauna, aquatic, terrestrial and socio-economical) during the construction and operation phases of the project must be briefly discussed for each alternative and the reasons be given for selecting the best and optimum alternative based on social and environmental considerations, and for rejecting other alternatives.

2.5.3 Inter-State Actions

One of the important regulations in water framework directives has to do with the spatial orientation towards the bodies of water. The reason is that one river may be shared by two or more countries. However, the country below the other one shall not be disadvantaged, e.g. by polluted river water caused by the geographically higher country. Thus, a common regulation in adjoining states obviously makes sense.

In Europe, as an example, the water framework directive, existing since 2000, is the guideline issued by the European Union concerning the treatment of water in the member states. The purpose is to form a common water policy for an environmentally sustainable usage of water. This may, e.g., concern the quality of the water as well as other aspects that are treated as following. If a river flows through different countries, an agreement is needed for operating a hydropower plant.

To emphasize the importance of this issue, the approach of India, as per the 'Hydro Policy 2005' is explained below as an example:

A substantial hydropower potential has remained locked up and many hydro projects could not be taken up for implementation, even though these projects are well recognized as being attractive and viable, because of unresolved Inter-State issues. The Government of India recognizes the need for evolving an approach to ensure that the available hydroelectric potential is fully utilized without prejudice to the rights of the riparian States as determined by the Awards of the Tribunals/ Agreements arrived at among the party States for a given river basin with regard to water sharing. The selection and design of the projects are based on integrated basin wise studies, so as to arrive at an optimal decision and care is taken that such projects do not in any way prejudice the claims of basin states or affect benefits from the existing projects. A consensus is evolved amongst the basin states regarding the location of such projects, the basic parameters involved and the mechanism through which each projects will be constructed and operated. As far as possible, preference is given to take up simple run-of-river schemes that do not involve any major storage or consumptive uses.

2.5.4 Joint Venture

Hydropower plants require a lot of investment. In many developed as well as developing countries, private companies or state owned companies are able to finance power station on their own. But sometimes local governments do not have sufficient funds, especially in developing countries. In such situations, private investors are encouraged to become partners with governments to form joint ventures for hydropower plants. In these cases the interaction of the partners can be manifold. Given below is an example from India that explains this mechanism.

When it comes to procuring additional private investment in India, schemes through joint ventures between the Public Sector Undertakings (PSU)/State Electricity Boards (SEB) and domestic and foreign private enterprises are preferred. If a joint venture company is created, it is an independent legal entity registered under the Companies Act and acts as an independent developer. The joint venture agreement between the two partners states clearly the extent of participation of each partner and the risks to be shared in relation to the implementation and operation of the project. The agreement also provides for arrangements in such cases where the joint venture partner will not be associated with the operation and maintenance of the project. While the selection of a joint venture partner is made in accordance with the policy of the Government, there is an option for the PSU to either select the joint venture partner together with their financial and equipment package or to select a joint venture partner wherein the Energy Performance Contract is decided by both the partners after they have formed the joint venture company. The associated transmission line connected with the scheme is constructed by the Powergrid Corporation of India. The power from joint venture hydel projects will be purchased by the Power Trading Corporation proposed to be formed with equity participation from Government/ Central Government PSU/Financial Institutions. The security for payment of power purchased from the joint venture projects is provided through a Letter of Credit to be provided by the SEB and recourse to the State's share of Central Plan

Table 2.2 List of the three largest hydropower plants in India that have reaped CDM benefits (as of May 2010)

UNFCCC Ref. No.	Title	State	MW capacity
862	Allain Duhangan Hydroelectric project (ADHP)	Himachal Pradesh	192
1326	Jorethang Loop Hydroelectric project	Sikkim	96
1844	Budhil Hydro Electric Project, India (BHEP)	Himachal Pradesh	70

Allocation and other devolution. This security package enables to raise finances for these projects. The State Government (home State/States) will be compensated by way of 12% free power as per the present policy applicable for Central Sector hydel projects in India (see also [1]).

2.5.5 *Land Acquisition, Resettlement and Rehabilitation*

The acquisition of requisite Government, forest and private land involves cumbersome procedures and difficult negotiations with land owners who have to part with their land. Demands for employment in lieu of the land cost, land for land at places of land owners choice etc. result in contractual problems for several projects. All such costs incurred by the developer are considered as costs of the project and allowed to be passed through tariff.

2.6 Clean Development Mechanism: Example of India

As per the guidelines of the United Nations Framework Convention on Climate (UNFCCC), small as well as large hydro power plants are eligible for being considered as candidates for Clean Development Mechanism (CDM) and Joint Implementation under the agreement for reducing greenhouse emissions as per the Kyoto Protocol. Due to the differences in environmental implications, a project is considered to be a small scale project, if the capacity of the hydro plant is up to 15 MW, and it would be subject to the approved methodology AMS I D. If the project capacity is >15 MW, the project will be considered to be a large scale project and be subject to the approved consolidated methodology ACM0002. If, in the case of storage type hydropower plants, the project activity is implemented in an existing reservoir where its volume is increased or if the project activity results in new reservoirs, the power density of the power plant must be greater than 4 W/m^2, and the project should comply with the World Commission on Dams (WCD) guidelines.

Table 2.3 List of three smallest hydropower plants in India that have taken CDM benefits (as on May 2010)

UNFCCC Ref. No.	Title	State	MW capacity
1566	Mini Hydel Scheme on Nagavali River, Andhra Pradesh	Andhra Pradesh	1.7
662	Link Canal Mini Hydel project	Karnataka	1.5
1512	Deogad hydroelectric project in Maharashtra district Sindudurg, India by M/s Gadre Marine Export	Maharashtra	1.5

As this an opportunity of additional earning by selling carbon credits, there has been a significant growth of the small hydro sector in India over past few years. A total of 64 Indian projects with a total installed capacity of about 1,100 MW have been registered with UNFCCC for capturing CDM benefits (see also Tables 2.2 and 2.3).

References

1. a) National policy for hydropower development, Report from the national hydropower corporation, Government of India. b) Policy on hydropower development, Policy document from Ministry of Power, Government of India, 1998. c) Hydro development plan for 12th five year plan, Central electricity authority, Government of India, New Delhi, 2008 http://www.nhpcindia.com/writereaddata/English/PDF/hydro-policy.pdf

Chapter 3
Physical and Technical Basics of Hydropower

3.1 Locating a Hydropower Plant

One of the most important factors governing the amount of power generation and the performance of a hydropower plant is its location. There are several considerations for identifying the most appropriate location for any large and small hydro power plant.

3.1.1 Considerations for Quantity of Water

When deciding on a location for a hydropower plant, it is necessary to know the average expected supply of water available for power generation. Since the amount of rain fall varies from year to year, and cannot be predicted very accurately, the decision about water availability is to be based on the historical data of previous decades. But still there is no guarantee for an annual average volume of water being available.

Due to variation in rainfall and change in temperature resulting in varying amounts of snow melting, the discharge into rivers varies from day to day. Therefore, a river power plant does not operate continuously at a constant rate. The expected daily flows during a year change quite randomly and are presented in the form of hydrographs for the duration of one year. To calculate the amount of energy expected to be produced during a year, the mean daily flows have to be sorted according to discharge of water into the river. This leads to a flow-duration curve, showing the period of time within a year in which a certain flow is reached or exceeded as shown in Fig. 3.1.

In addition to the flow-duration curve, some other curves have to be considered, such as the mean duration curve of the water surface elevation in the upper reservoir depending on water input and output by down streaming water.

H.-J. Wagner and J. Mathur, *Introduction to Hydro Energy Systems*, Green Energy and Technology, DOI: 10.1007/978-3-642-20709-9_3,

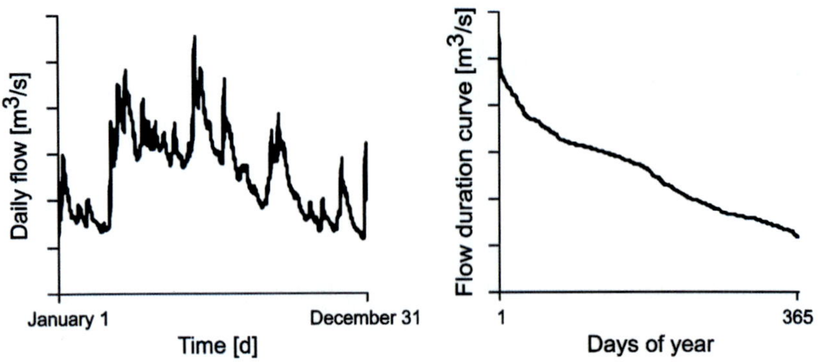

Fig. 3.1 Hydrograph and flow duration curve

When this information is taken into account the yearly energy output can be calculated by integrating the water power equation over the year. A good estimation of the expected mean power generation per year is only possible if long-term historical stream flow records are available, usually of 20 years or more. By using stochastic hydrology with correlation techniques, short records may be extended to a longer suitable duration.

When constructing a hydroelectric power plant, it is necessary to know the expected average availability of water. Since the amount of rain is not static and cannot be predicted, the empirical equations based upon the data of previous decades are used. However, there is no guarantee for the estimated availability of water. In the case of the Rhine River, the amount of water of the Rhine at Rheinfelden in the years 1999, 2001, and 2003, and the flow rates were registered daily, as shown in Fig. 3.2. The Rhine has an annual average volume of water of about 1,000 m^3/s. While in the summer months the average supply amounts up to 1,500 m^3/s and above, the winter months deliver around 800 m^3/s of water. The year 2003 was a relatively dry year, and 1999 and 2001 were relatively wet years. In the summer months of 2003, when the weather is usually wet, the flow rate was only between 1,000 and 1,200 m^3/s. The months of January and October, on the other hand, showed even higher flow rates than the summer months. In 1999 there were two very wet periods, one in February and March and the other one from May to July. Then, the flow rates accounted to more than 4,000 m^3/s, which is a very high amount. The year 2001 was not as wet as 1999 but still shows high flow rates from all of March to August. For hydroelectric power plants oriented to the average volume this meant that in 2003 the installed capacity was too high for the actual supply of water and so the plants were not working to their full capacity. In the year 1999, however, the capacities of the power plants were not high enough to be able to use the whole water amount and part of the potential energy was lost, e.g. by flooding over the dam.

In the light of this example, how much sense does it make to construct a power plant based on the average water supply? If a power plant is not able to produce electricity due to a shortage of water, money is wasted for providing redundant

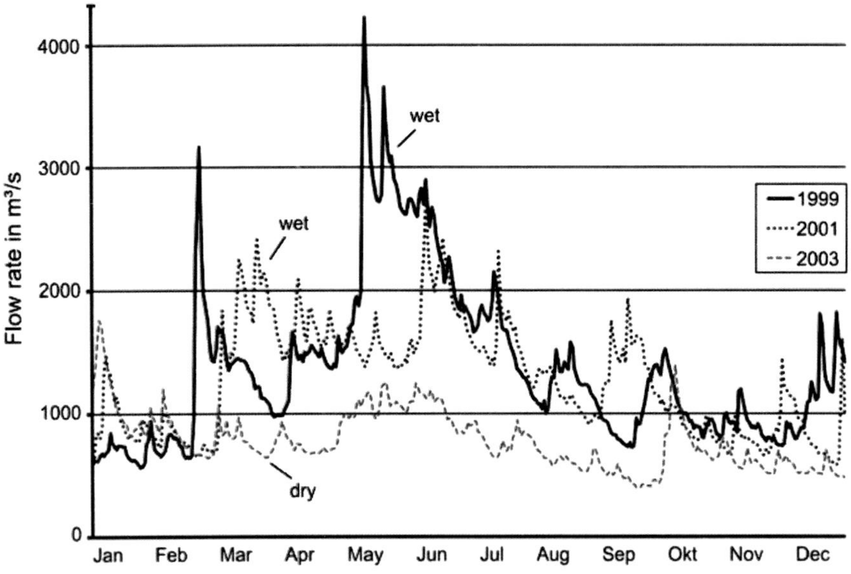

Fig. 3.2 Hydrograph of the Rhine River in Germany of two selected "Wet" years (1999 and 2001) and one "Dry" year (2003) [1]

capacity. On the other hand, the availability of surplus water is an opportunity lost to generate more power that could bring more revenue. The final decision about capacity therefore depends upon the availability of water together with the interest rate, market price of electricity, and the resulting payback period.

Another specification that needs to be derived from the availability of water is the firm power or secure power. This entails an estimation of that amount of electricity that will be surely available throughout the year. For estimation of secure power, the claimed amount of power should be available for a certain period over one year. In practice, however, the river flow rate may be higher or lower than the turbine flow rate, i.e. the flow required by the turbine to produce maximum power as per its capacity. When the flow is lower, the turbine uses the stored water in the reservoir. When the river flow rate is higher than the turbine flow rate, the reservoir is filled with extra water. However, when the reservoir is completely filled up, and more surplus water is expected to come, the excess water is passed forward through the spillway of the dam, bypassing the turbine.

3.1.2 Considerations for Location of Hydropower Plant

Depending on the topographic situation at the river-site where a hydropower plant is planned, there are several possibilities to locate the power house. One possible layout is to place the power house into an artificial bay at one of the river banks.

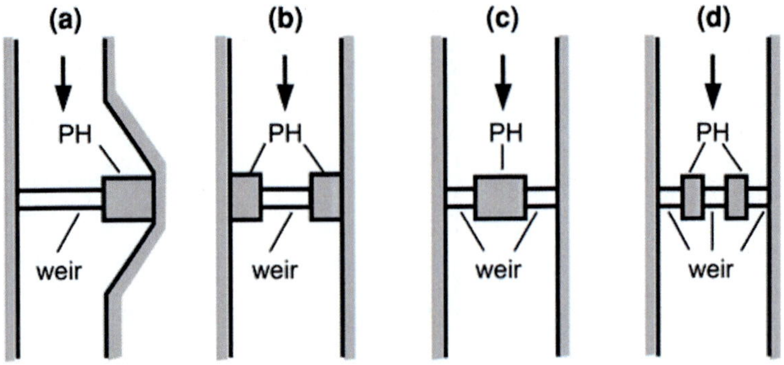

Fig. 3.3 Layout of power plant and weir in the River Bed (PH = Power House)

This is the construction of choice when the facility including all auxiliary structure (weir, sluice) is too large to fit between the natural river banks. These types of power plants are called 'bay power plants'. Various layouts are shown in Fig. 3.3(a-d). The most important advantage of layout (a) is the optimal passage of floods at the weir adjacent to the power house. If circumstances permit and the width of the river is large enough all buildings including the weir, are placed into the original river bed without affecting the flood passage. In this case, one has several possibilities to place the powerhouse relating to the weir. Especially when a power plant is planned by two countries as a joint venture and is located at a border river, option (b) is suitable. In this case, each country can run its own plant without interfering in the operation of its neighbour. Due to the inconvenient accessibility of the power house, layout (c) is only recommended when foundation conditions call for this location. A widespread arrangement is option (d) where every single turbine along with a coupled generator is placed into a separate pier of the weir. In this case the units are accessible by a bridge crossing the river.

Ecological and landscape aspects sometimes lead to designs differing from the above described construction methods (Fig. 3.4).

In river bends the power station is located at the outside curve because of the so-called spiral flow. In rivers with sediment transport, this rotating current leads to erosion of the outside curve and to silting of the inside parts of the river bed as shown in Fig. 3.5. Hence, to protect the intake structure of the power house from silting, it has to be placed at the outer river bank.

As an option, in many cases, a diversion power plant (see also Chap. 1) may be considered which consists of a canal before the weir taking water to the power plant, which leaves the main course of the river undisturbed. The advantage of this type is that other activities such as transportation by ships, are not disturbed by the presence of the hydropower plant. However, if the quantity of water in the river is not large, the main river may not be left with sufficient water after most of the water has been diverted to the canal. This situation causes problems for those activities that earlier were dependent upon the river water. However, after the discharge from the power plant has again joined the main course, there is sufficient flow again.

Fig. 3.4 One hundred year old river power plant in Black-Forest Region in Germany, *Left Side* is power house with water channel, *Right Side* the arched structure of weir

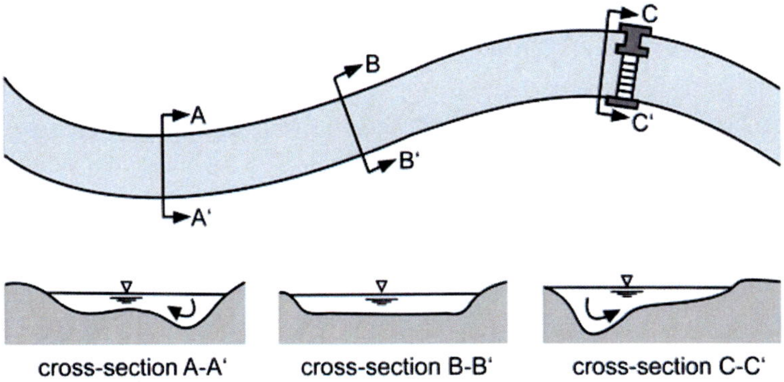

Fig. 3.5 Development of spiral flow and silting in River Bed

3.1.3 Multiple Reservoir System

In some cases, especially in mountainous areas, instead of only one large catchment area, there may be several smaller catchment areas providing for more than one reservoir. This situation is shown in Fig. 3.6. Water from these scattered reservoirs cannot be collected to make one single large reservoir due to geographic

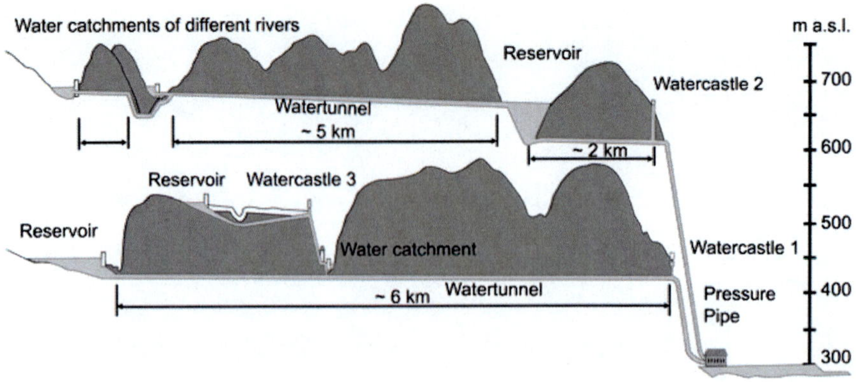

Fig. 3.6 Multiple reservoir system of a storage power plant in Black-Forest region, Germany (m.a.s.l. = *m*eter *a*bove *s*ea *l*evel)

limitations, and the installation of separate power plants for each small reservoir may not be possible due to the lack of a suitable place for installing the turbine. Figure 3.6 shows that the location on the right hand side is the only suitable place for installing the turbine, since it offers the largest head for power generation. In such cases, water is brought to one common power plant through water tunnels running on the ground as well as through trenches dug through the mountains. With this arrangement, the turbine and the reservoirs may even be located several kilometres apart and may not even be visible from each other.

3.1.4 Cascaded Hydropower Plants

In certain situations, a large quantity of water is available and a good head of water is available over a large horizontal distance. In the case of this combination, where the quantity of water has to be transported over long distances, a common power plant might not be the best choice. Therefore, to fully capitalize on the fall height and quantity of water, a series of river power plants is created. A schematic of such an arrangement is given in Fig. 3.7. It is important to operate these power plants as river power plants and not storage type power plants, since water discharge from one affect the power output from all subsequent power plants in series.

3.2 Basics of Fluid Mechanics

3.2.1 Characteristic of Water

Water, the molecule consisting of one oxygen and two hydrogen atoms, has a number of special features that are important for both construction and operation of hydropower plants.

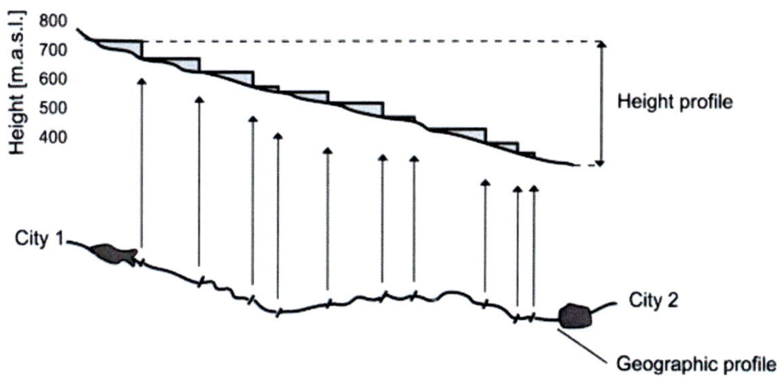

Fig. 3.7 Scheme of cascaded hydropower plants on a River (m.a.s.l. = meter above sea level)

One characteristic of water is the density of one kilogram per litre under normal conditions, i.e. a temperature of 20°C and a pressure of 1 bar. Another property is the incompressibility of water in fluid state. When the pressure of water is increased, it transmits the pressure to its surroundings; water cannot store the effect of an increase in pressure by getting compressed as is the in case with gases. For hydropower plants this means that a sudden increase of pressure may disturb the operation and even destroy machine parts. In order to avoid damages, various measures are taken to avert a sudden transmission of pressure from machine parts of the power plant.

Water shows an increase of volume if its temperature decreases below +4°C. The maximal volume is reached at –4°C. Due to this property, water expands and may cause damage to machine parts or tear up the soil. Therefore, in cold regions, the freezing of water must be prevented by suitable heating systems, e.g. by a warming grill.

The evaporation temperature of water, on the other hand, depends much on the pressure. With a pressure of 1 bar water evaporates at 100°C, whereas a pressure of 20 mbar makes water evaporate at only 18°C. This phenomenon is the reason of the so-called cavitation, which will be described in detail in Sect. 3.2.6.

Further, water enters into solutions with acids, leaches, and salts. Due to this feature damage may be caused, e.g. corrosion of machine parts.

3.2.2 Velocity Equation

Running water, as in rivers, transports sand and gravel due to the high energy it contains. Sand starts moving with water from a speed of about 0.3 m/s, smaller gravel from 1.0 m/s and bigger gravel from about 1.3 m/s. These particles move with the water until the speed of the water slows down, either at the bends or after reaching the reservoir. The effect of this transportation is that sand and gravel accumulate in a pile in front of the dam wall restrict the operation of the power plant or cause other problems, particularly with the function of sluices for ships.

Energy is stored in water due to its elevation or fall height. When water falls over a height h, the potential energy is converted into equivalent kinetic energy as shown in the equation below:

Loss in potential energy = Kinetic energy after falling over fall height h.

$$mgh = \frac{1}{2} \cdot m \cdot v^2 = \frac{1}{2} \rho \cdot V \cdot v^2 \tag{3.1}$$

where:

m	mass of water falling
v	velocity
g	acceleration due to gravity (9.81 m/s^2)
h	height
ρ	density of water
V	volume

and the above equation can be rewritten to give the velocity of water falling over a height h:

$$v = \sqrt{2 \cdot g \cdot h} \tag{3.2}$$

This equation suggests that the velocity increases with square-root of height of fall. To give an impression on the energy content of water the potential energy of one cubic meter water at a level of 100 m contains the energy of 0.25 kWh.

3.2.3 Bernoulli's Equation

The water flowing in the river and the water in storage possess two type of energy: the kinetic energy due to the water's flow and the potential energy due to the water's height. In hydroelectric power plants the turbines are driven by kinetic energy.

Like all other cases related to energy, water flow also follows the law of energy conservation. As per the Bernoulli's theorem, for a non-viscous, incompressible fluid in steady flow, the sum of pressure, potential and kinetic energies per unit volume is constant at any point. Bernoulli equation is a special form of the Euler's equation derived along a fluid flow streamline, and can be expressed, as shown below, by three terms:

kinetic energy + pressure energy + potential energy = constant

$$\frac{v^2}{2} + \frac{p}{\rho} + gh = constant \tag{3.3}$$

where:

v	flow speed
p	pressure

ρ density of water
g acceleration due to gravity (9.81 m/s^2)
h height of fall

If water is only stored at a fall height h in an open reservoir, the Eq. 3.3, has only one component for potential energy. The pressure head and the kinetic head are absent.

3.2.4 Power Equation

Hence, the total power that can be generated from water in hydroelectric power plant due to its height is given by:

$$P = \eta \cdot \dot{m} \cdot g \cdot h = \eta \cdot \rho \cdot Q \cdot g \cdot h \tag{3.4}$$

where:
P total power that can be produced
$\dot{m}$ mass flow of water falling $= Q \cdot \rho$
η overall efficiency of power stations
ρ density of water
Q flow rate of water = Volume V per unit time
g acceleration due to gravity (9.81 m/s^2)
h height of fall.

The above equation shows that the power output from a hydropower plant is proportional to two natural parameters, i.e. volume of flow and height of fall. The next important parameter is overall efficiency, which can be improved through proper selection and operation of machinery. The overall efficiency η is by neglecting the losses in pipes the product of the turbine efficiency and the generator efficiency. The turbine efficiency usually ranges between 0.85 and 0.95, depending on the type and design of the turbine used, and takes into account efficiency losses due to friction and turbulence between the entrance of the turbine and the end of the draft-tube. Friction losses within the generator lead to heat and noise in the machinery and powerhouse, and are included in the generator efficiency of about 98%. The overall station efficiency can be raised by increasing the number of installed units, especially when flows are fluctuating.

In order to obtain a high head of water, the water reservoir should be situated as high as possible, and the power generation unit should be located as low as possible. The maximum height of a water reservoir is determined by geographical factors, such as the height of the river bed, the amount of water and other environmental factors. The location of the power generation unit can be adjusted as per the total amount of power that is to be generated. However, the location of the power generation unit is also subject to geographical constraints since usually the power generation unit is installed at levels lower than the local ground level so as to get the maximum head of water.

The total flow rate of water can be adjusted through an important part of the hydropower station: the penstock. Its dimension must be optimized for the rated power of the hydro power stations.

Power Capacity in Hydropower Plants

With the help of a simple hydropower plant, Eq. 3.4 shall serve an example. Assumed were the below mentioned parameters:

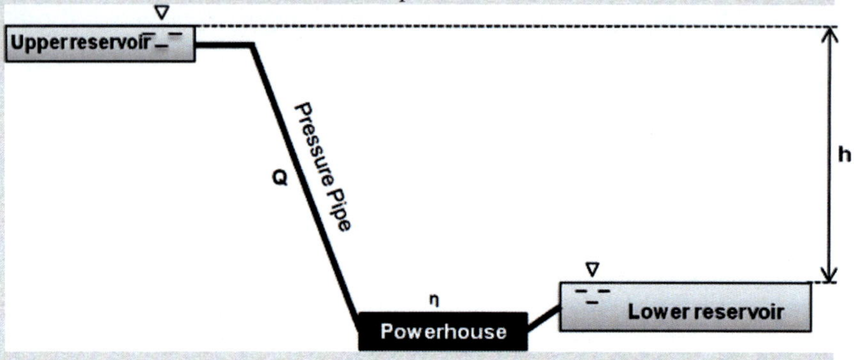

Height of fall h: 100 m
Gravity constant g: 9.81 m/s^2
Density of water ρ: 1,000 kg/m^3
Flow rate of water Q: 50 m^3/s
The overall efficiency of hydropower station η: 89 %
This leads to a total power output of:

$$P = 0.89 * 1,000\ \mathrm{kg/m^3} * 50\ \mathrm{m^3/s} * 9.81\ \mathrm{m/s^2} * 100\ \mathrm{m}$$

$$P \approx 43.7 * 10^6 \frac{\mathrm{kg\,m^2}}{\mathrm{s^3}} = 43.7 \frac{\mathrm{MJ}}{\mathrm{s}} = 43.7\ \mathrm{MW}$$

3.2.5 Continuity Equation

The continuity equation of water suggests that due its incompressible nature, the product of velocity of flow and cross section area of flow remains constant. The continuity equation for water is given by:

$$v_1 A_1 = v_2 A_2 = v_3 A_3 = \ldots = \text{constant} \tag{3.5}$$

where:
v_1, v_2, v_3 are velocities at three different sections having area A_1, A_2, A_3.

Table 3.1 Evaporation Temperature of Water at Different Pressures

Temperature (°C)	Pressure (bar)
2	0.0071
4	0.0081
6	0.0093
8	0.0107
10	0.0123
12	0.0140
14	0.0160
16	0.0182
18	0.0206
20	0.0234
22	0.0264
24	0.0300

This equation shows that with a variation of the cross section of flow, a change in the flow velocity must occur in order to satisfy the continuity equation. For a smaller cross section, the flow velocity increases and for larger ones, it decreases.

3.2.6 Cavitations

Considering the effect of the continuity equation together with Bernoulli's equation suggests that with a decrease in cross section, there is an increase in flow velocity that increases the kinetic head of fluid. Since the total of all three terms in Bernoulli's equation has to remain constant, this increase in kinetic head results in the reduction in pressure if the level of water does not significantly change.

Water also has the property of boiling at a low temperature with a decrease in pressure, and if this static pressure decreases significantly, it leads to evaporation of water. The bubbles of water vapour formed due to evaporation, are transported towards the low pressure zone where they collapse causing a rush of water to fill the empty space of the bubbles. If this phenomenon, called 'cavitation' takes place near surfaces of equipment such as turbine blades, is so strong that it causes permanent damage to parts besides causing a loss of efficiency.

Therefore, technically speaking, cavitation can be defined as formation of voids within a body of moving liquid when the particles of liquid fail to adhere to the boundary of the passage way. Failure of the particles to adhere to the boundaries occurs when there is insufficient pressure to overcome the inertia of the particles and to force them to take the path according to the curvature of the path of flow. Since the inertia of the particles varies with square of their velocity, the phenomenon of cavitation is usually associated with cases of high velocity and low pressure flow conditions.

Water, under normal conditions, i.e. at a temperature of 20°C and a pressure of one bar, is fluid. With the same pressure it evaporates at 100°C. When the pressure is reduced, water evaporates at lower temperatures and this is a problem with hydropower plants. With a pressure of 20 mbar, water evaporates at only 18°C as can be seen from Table 3.1.

Fig. 3.8 Damage of a rotor blade as a result of cavitation

In storage power plants, for instance, water must be sucked in from the lower level to be pumped up to the upper water level. At that moment, when water is sucked in, its surrounding pressure is reduced. The consequence is that the water evaporates at low temperatures, at low pressure points. After being sucked in the water reaches areas at the pumping blades and in the pipes the pressure rises again. Due to high pressure, the vapour of water condenses back and the water becomes fluid again, in the form of tiny drops. These drops have a very high velocity with which they smash machine parts, like the pump or the turbine. Cavitation does not only appear when water is sucked in but also with water in pipes. The pressure in the pipes depends on the velocity of water, and the velocity depends on the cross-section of the pipe. Under these circumstances, the inner flow pressure decreases and cavitation may emerge.

This can cause material damage, like dents looking like little holes in the surface of the turbine blades as shown in Fig. 3.8, which evoke higher friction in the running of the turbine. Also the mechanical stability of the blade could be endangered. The rotor can be repaired by welding up the holes and smoothing the welded spots. However, apart from the fact that the operation of the power plant has to be interrupted, the repair does not last for long, if the problem of cavitation persists.

Cavitation can be deceased or eliminated by increasing the pressure in the discharge side of the turbine blades or by decreasing the velocity of flow.

The pressure on the discharge side of the turbine can be increased by locating the turbine runner lower than the tail-race. It is not always possible or economical to do so, since there is a loss of net head. The other option of decreasing the velocity of flow is achieved through increasing the diameter of pipes. In order to avoid cavitations in pipes, they are constructed with a wide diameter. The diameter is ideally constructed with respect to the calculated velocity of water. The limitation with increasing the discharge pipe diameter is that it also requires to increase the size of the turbine runner, which in turn increases the cost of the turbine.

Modern turbines are very carefully designed with the help of simulation tools based on Computational Fluid Dynamics (CFD) based simulation tools to avoid too many cavitations.

However, it is not possible to avoid cavitations totally in the daily operation of hydropower stations.

Reference

1. Bundesamt für Umwelt: http://www.hydrodaten.admin.ch/d/2091.htm

Chapter 4
Components of Hydropower Plants

4.1 Main Parts

In the following are the main parts of any hydropower plant that are needed to convert the energy in water into electricity. Without these parts, generation of power by a hydropower plant is nearly impossible.

4.1.1 Turbine

The turbine can be considered as the heart of any hydropower plant. Its role is to convert the power of water into mechanical power, i.e. by rotating the shaft. Hydraulic turbines have a row of blades fitted to the rotating shaft or a rotating plate. Flowing water, while passing through the hydraulic turbine, strikes the blades of the turbine and makes the shaft rotate due to its impact or change of velocity and pressure. While flowing through the hydraulic turbine the velocity and pressure of water diminish resulting in the development of torque and rotation of the turbine shaft. There are different forms or designs of hydraulic turbines in use depending on the operational requirements. The selection of the turbine is very critical for the success of a hydropower plant. The optimum output of a specific combination of water flow and head can only be achieved by an adequate type of hydraulic turbine. Details about types, selection and operation of turbines are given in a separate chapter.

4.1.2 Electric Generator

Similarly to any other generator, the primary function of a generator for a hydropower station is to convert the rotation of shaft into electric power. Next to

H.-J. Wagner and J. Mathur, *Introduction to Hydro Energy Systems*,
Green Energy and Technology, DOI: 10.1007/978-3-642-20709-9_4,

Table 4.1 Synchronous generator speeds in records per minute (rpm)

No. of poles	No. of pole pairs	Generator speed for 50 Hz supply	Generator speed for 60 Hz supply
2	1	3,000	3,600
14	7	428	514
18	9	333	400
20	10	300	360
60	30	100	120

the turbine, the most important part of a hydropower plant is the generator. The basic process of generating electricity in this manner is to rotate a series of coils inside a magnetic field or vice versa. This process leads to the movement of electrons inside conductors, which produces electrical current.

The rotational speed of a synchronous generator which may be directly connected to a grid is kept constant, and dictated by the constant frequency of the grid. The relationship between the rotational speed of generator and frequency is governed by the formula:

$$\text{Rotation} = \frac{\text{Frequency (50 Hz or 60 Hz)}}{2 \cdot \text{No. of poles}} \tag{4.1}$$

The number of poles indicates the number of the sets of coils in the stator of the generator in which the electric power is generated. Two poles make one pole pair hence the number of poles in a generator cannot be an odd number. Due to the fact that grid is operating with three phases, the generator must also produce three phase alternating current. Only in very small plants that are not connected to the grid, the generators are of single phase, The difference is similar to the fact that most of large machines use three phase motors and only small machines use single phase motors.

When using this formula it must be taken into account that the popular unit of rotation is 'revolutions per minute' and the unit of frequency 'cycles per second'. Therefore, a conversion factor must be applied in the above equation to convert rotation and frequency into the same unit. Table 4.1 shows the change required speed of generators with change in the number of poles.

In practice, the speed of large hydro turbines is not normally more than 500 rpm, requiring a generator of around 8 or more pole pairs.

Turbine and generator of a hydropower plant are connected through a shaft. Figure 4.1 shows the shaft of a 70 MW machine. This turbine has 500 rpm rotational speed, and its generator has 6 pole pairs for producing 50 Hz current.

4.1.3 Transformer and Power House

The transformers and power house of a hydropower plant act as an interface between the electric generator and the power transmission lines. The voltage of the generated middle voltage electricity by generator, e.g. 6,000 V, is increased into

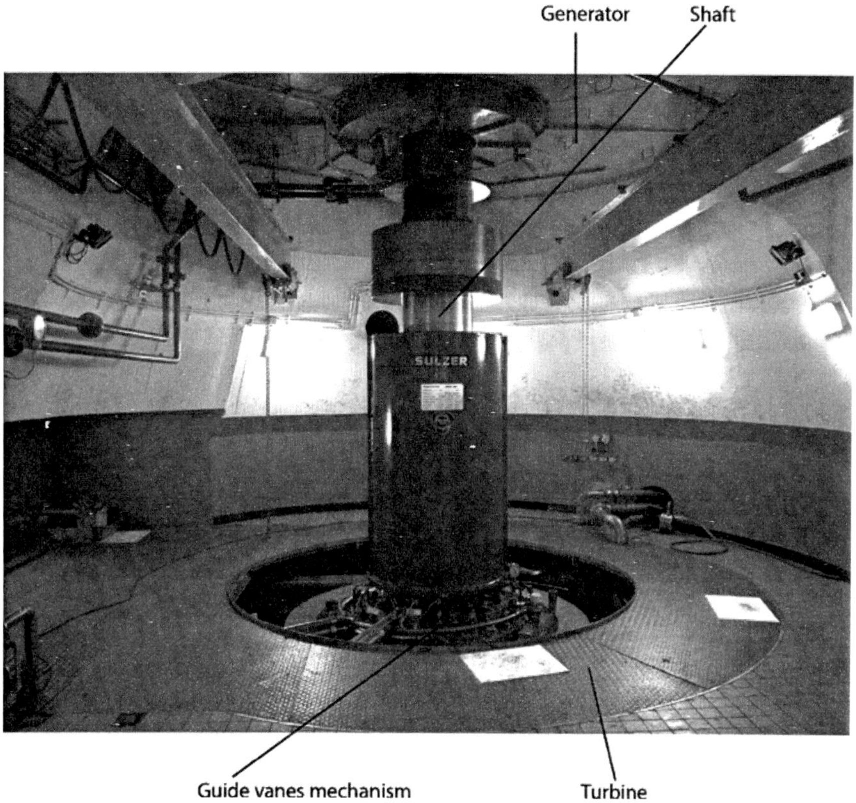

Fig. 4.1 Shaft of 70 MW hydropower plant in Roenkhausen, Germany

very high voltage electricity by using transformers. Such a conversion is required since high voltage is preferred for transmitting of power over long distances as technical losses are reduced.

4.1.4 Upper and Lower Reservoir

Storage type hydropower plants have an upper and a lower water level with the machinery in between. As the word 'storage' suggests, the water is stored in a reservoir before it comes in contact with the turbine. Water is stored in upper reservoir because during the course of the day or with seasonal changes, different amounts of electricity are demanded from the grid and the availability of water in the river feeding water to the hydropower plant may not match the requirement of power governing the requirement of water for the turbine. Such reservoirs may be naturally available close to the site of power generation; however, in most cases they are man-made reservoirs only.

Fig. 4.2 Sludge deposited in reservoir of 1,000 MW storage power plant Vianden, Luxemburg

Some storage power plants have another reservoir at the discharge end of the turbine. Such discharge side reservoirs are found in pumped storage power plants. In such power plants, the water stored in the upper reservoir is first converted into electricity when streaming down through the turbine. This water is stored in the lower reservoir and is later pumped up to the upper reservoir to have potential energy again, when surplus electricity is available in non-peak hours. The lower reservoir is usually smaller than the upper reservoir that primarily depends upon factors such as availability of water and power demand peaks in the load profile, besides feasibility related aspects such as availability of a suitable place for locating a lower reservoir. Figure 4.2 shows a photograph of an emptied upper reservoir of a 1,000 MW pumped storage power plant in Luxemburg. The photograph shows the amount of sludge that gets deposited at the bottom of the reservoir which is to be cleaned periodically.

It may be noted that sometimes, due to the non-availability of a natural reservoir, an artificial reservoir needs to be created.

4.2 Structural Parts

Structural parts of hydropower station are those parts that do not directly take part in power generation; however, they form the basic structure that facilitates controlled and safe use of water for power generation by turbine and generator.

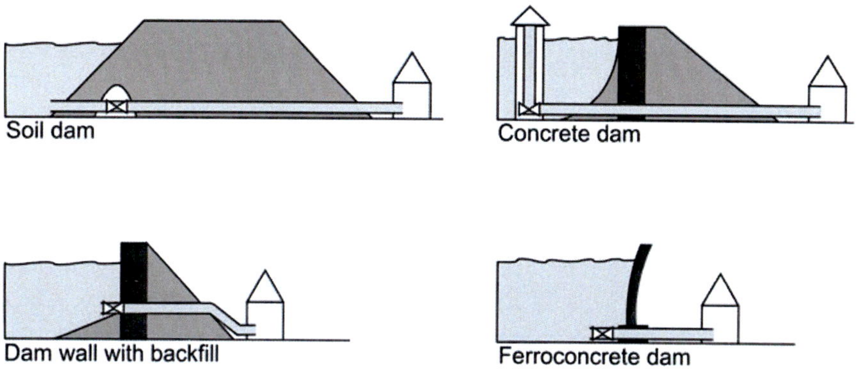

Fig. 4.3 Four types of dam walls

In the following the parts are described that constitute the main structure of the hydropower plant:

4.2.1 Dam and Spillway

Dams are structures built over rivers to stop the water flow and form a reservoir. The reservoir stores the water flowing down the river as explained in the previous section. This water is diverted to turbines of hydro power stations. The dams collect water during the rainy season and store it, thus allowing for a steady flow through the turbines throughout the year. Dams are also used for controlling floods and to store water for irrigation purpose. The prime requirement for any dam is to be able to withstand the pressure exerted by the huge amount of water that is stored behind it. There are different types of dams depending upon the shape of their structure such as arch dams, gravity dams and buttress dams. The height of water in the dam is called 'head race'.

For creating a dam, the river needs to be diverted from the place of construction, and especially must be drained so that construction actively can take place. This is realised by a temporary reservoir surrounded by metal walls. The reservoir area is kept dry and the construction is quickly carried out. As the structure of a dam spreads, sometimes it is required to shift the temporary reservoir according to the construction plan and inflow of water.

There are different types of dam walls, four of which are shown in Fig. 4.3. The soil dam is to be mentioned as it is one of the earliest types (from eighteenth century). It consists of a trapezoid bank of soil and/or stones towards upper and lower water level. The dam is covered with an isolation of flagstones of clay and/or bitumen.

Another type is the concrete dam. It is similar to the former type, but contains a concrete core and a concrete pin which provides more stability to the structure. The pin is positioned in the middle of the dam and reaches into the solid ground.

This version of dam was first constructed around 1850. A third dam type is a dam with backfill. This type does not have the trapezoid form anymore. The wall side facing the upper water is straight, while the backfill is on the lower water side.

The fourth type of dam walls shown in Fig. 4.3 is the ferroconcrete dam. This is the most recent constructional type which is used for the most colossal dams, like huge dams for storage power plants. Here the dam simply consists, as the name indicates, of ferroconcrete and does not need additional backfill. The exact form of the wall is bowed to a certain, calculated degree of bend or angle which guarantees the highest stability (Fig. 4.4). Every dam has an outlet pipe with a valve which could be positioned at different places.

Spillway is an integral part of any dam. A spillway as the name suggests is a method for spilling of water from dams. It is used to provide for the release of flood water from a dam. It is used to prevent over toping of the dams which could result in damage or failure of dams. Spillways are of two types: controlled type and uncontrolled type. The uncontrolled type of spillway is one which starts releasing water when water rises above a particular level. In this type, overflow is the only way for water to reach the other side of the dam. In the case of the controlled type spillway, it is possible to regulate flow through gates provided within the dam structure that provide an opening for releasing water to the other side without passing it through the turbine. Figure 4.5 shows the 60 m high spillway of a 100 year old hydro power plant in Schwarzenbach, Germany.

The creating a dam leads definitely to the formation of a huge reservoir in front of the dam that expands far beyond the former riverbank. As a result of the spread of water, dry places become submerged under water. This intervention in the surrounding environment changes the local nature and has an impact on topographic habitats. Due to this, the environmental impact assessment provides for compensatory measures. Examples for these are the building of a pond for frogs, reforestation, or constructing wet meadows and other biotopes.

4.2.2 Surge Chambers

These are tower like structures that provide chambers for the temporary storage of water. The role of surge chamber is to provide buffer space for the storage or supply or water in case of sudden increase or decrease in turbine loading. This is a construction before the inlet to the turbine that has regulatory functions. When the valve of a hydro power plant supplying water to the turbine is opened, it may happen that the water head is interrupted. In such a situation, the pressure of the water in the supply pie gets reduced and the possibility of cavitations arises.

On the other hand, when the valve is shut, due to a sudden stoppage of incoming water, the pressure in the supply pipes rises suddenly and the pipes may even burst. The surge chamber (see Fig. 4.6), which is a cylindrical water storage

Fig. 4.4 Dam wall of hoover hydropower station, USA (photograph by courtesy of Pamela Mc Creight, Wikepedia commons, licensed under creativecommons-Licence by cc-by-sa-2.0-de)

structure, is connected to the pipe between dam and turbine and regulates the swaying of water. When the valve is opened, the surge chamber can feed the water it contains to the pipes to avoid interruption of the water head. After the effect of a sudden change in flow has stabilized, the chamber is filled with water again to the

Fig. 4.5 Dam wall with spillway of 100 year old storage power plant in Black-Forest region, Germany (see also Fig. 4.9)

same level of the upper water reservoir. When the valve is shut again it regulates the increasing pressure by letting water in its empty space above the previous water level. Here the kinetic energy of water is converted into potential energy, which is a higher level of water. Once the flow stabilizes, the high level water automatically comes down to the level of the reservoir.

The practice and mechanism of operation of surge chambers of hydropower plants is principally the same as for surge chambers used in large water pumps. The role of surge chambers in each case is to prevent damage of equipment and structure against the effect of change in equipment loading.

4.2.3 Stilling Basins

Dam constructions have a further important structure that handles flooding the stilling basin. Flooding happens now and again and the power plant is not aligned

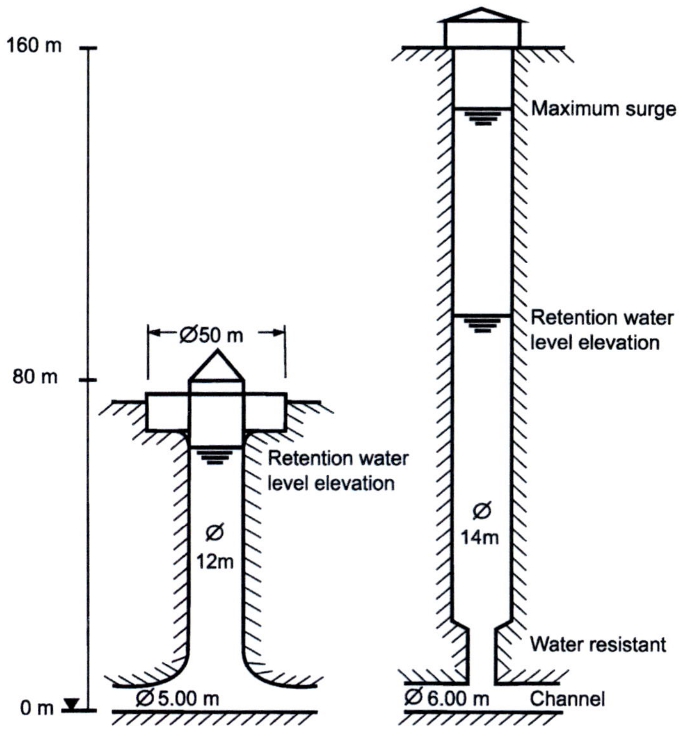

Fig. 4.6 Surge chambers of the Schluchsee storage power plant, Germany (concept of the figure taken by courtesy of König/Jehle [1])

to every possible water level and therefore the plant sometimes cannot reduce the overall amount of water to the regular level. Thus, as a part of the flood water cannot cross the turbine, it must exit the reservoir from somewhere else, e.g. over the dam into a small reservoir called stilling basin. The stilling basin's role is somewhat related to the spillway. While passing over the spillway, depending upon the flow quantity, water can summon up huge powers and it can be hard to control the damage caused by it. The function of a stilling basin is to reduce that danger. This is a type of basin that has a certain form which directs the water into a calculable subtle way and makes it easier to control. There are various versions of stilling basins; three types of stilling basin are described here:

One possibility to control the flooding water is using a regular basin with a pillar in the middle (Fig. 4.7). The water has to stream around the pillar and knocks together behind the pillar. As there is no other way to make way, the water squirts and spills over. While this happens the water loses its dangerous power and than can be safely brought to the lower water of the river.

Another version of a stilling basin is by providing stairs like structures behind the dam. While streaming down the stairs the water reduces its power before

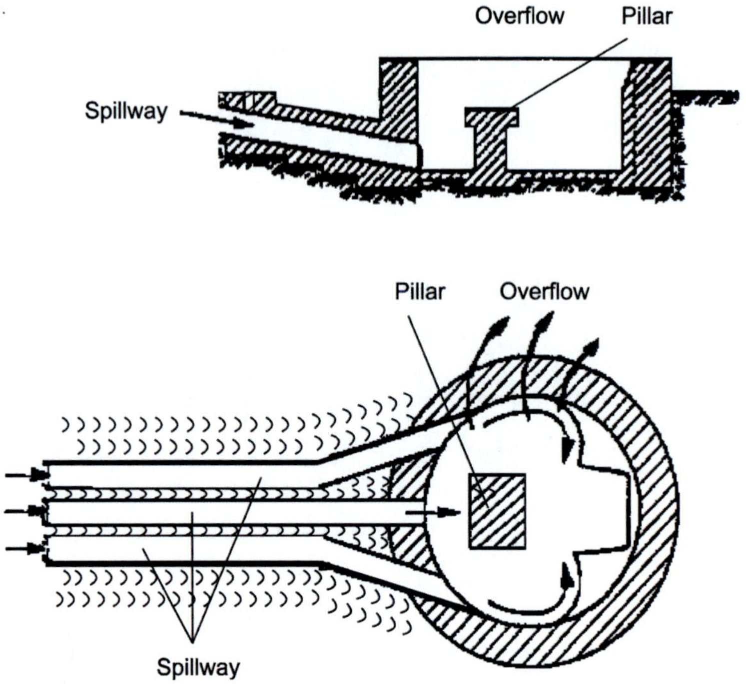

Fig. 4.7 Stilling basin with pillar (concept of the figure taken by courtesy of König/Jehle [1])

entering the river again. An important feature here is that the stairs are different in length and height. The reason for this construction is to avoid swapping of the water, which would not reduce the water's power as much.

A third type is a stilling basin with a sloping ground. When the water floods the dam it streams into the basin and spreads to each side. This movement enables the water to lose power.

4.2.4 Penstock and Spiral Casing

Penstocks are pipes which carry water from the reservoir to the turbines located inside the power station. They are usually made of steel or concrete, and are equipped with gate systems for controlling the flow. The pressure of water flowing through the penstock is very high. In some locations where an obstruction is present between the dam and power station such as a mountain, the tunnel connecting the reservoir and the power station itself serves the purpose of a penstock.

After passing through the screen or grill for preventing entry of large solid items such as stones or pebbles, the water faces the trumpet shaped inlet that is also called spiral casing as shown in Fig. 4.8. The cross section of the spiral casing

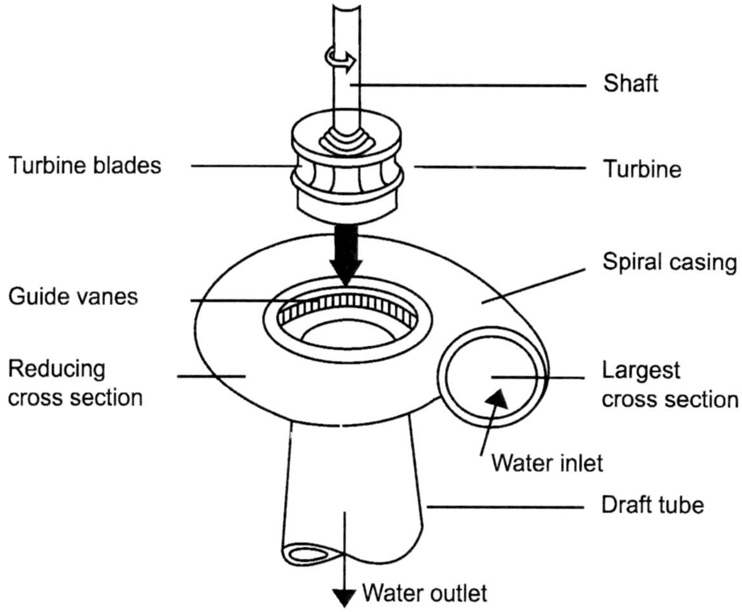

Fig. 4.8 Spiral casing of francis or kaplan turbines

constantly diminishes along its length circumferentially around the turbine. This part of the construction has the function to impinge water on the turbine symmetrically, which ensures that every part of the turbine receives the same amount of incoming water. If the spiral casing was absent, the turbine blades that receive water later would get less water than the part connected to the turbine that comes first in the path of the incoming water. The trumpet inlet can be in horizontal or vertical position and is constructed around Francis or Kaplan turbines. Furthermore they can either be made from steel or concrete or can be freestanding or built in the bottom.

4.2.5 Tailrace

The tailrace is the downstream part of a dam where the impounded water re-enters the river. It is the last part of the power plant structure, before the water enters the downstream river or lake. The tail race widens towards the end for the reason that with an enlarging section the energetic losses will be reduced as compared to losses in the constant section tail race. This loss of energy in outgoing water helps to reduce the back pressure on the upstream side and consequently helps in operating the turbine more efficiently. When the water is led to the lower water level in the outtake structure, the pressure of the lower water level makes it harder for the water coming from the turbine side to enter the downstream river. This happens under the surface of the lower water level. In order to reduce this problem the outlet structure is wider at its end to increase the pressure of the streaming water by slowing it down by use a widened section.

4.2.6 Pressure Pipes

Pressure pipes or pressure tunnels, as their name suggests, lead water under pressure and normally do not allow any air to enter the system. Their main purpose is to transfer water from the reservoir to the power plant. They are made by placing an additional layer of ferroconcrete between the layers of steel and rock. Its steel withstands the pressure in the pipe, while the concrete connects the pipe to the rock. Below the tunnel, separated from the actual pipe, there are additionally grooves for draining the pipe surrounding rock. Free level tunnels are more or less horizontal, while pressure pipes can be sloping or even vertical. The pressure pipes may have different tilt angles ranging from 30° to 90°. What is more important for pressure pipes and tunnels is the pressure head. At the bottom of a 500 m high pipe filled with water, the pressure would be 50 bar, for which they need to be built very strong sturdily. Figure 4.9 shows a photograph of 880 m long pressure pipes of a storage power plant in the Black-Forest region, Germany. The pipes are designed for flow of 6 m^3/s for supplying water to Pelton Turbines.

4.2.7 Caverns

Caverns, a structural part of any hydropower station, are constructed subterraneously within the mountain. They can even be built with their level below the lower water level in order to avoid cavitations, as explained earlier. There are different reasons for constructing a power plant within a mountain. At first there is the visual reason. The acceptance of a power station by nearby residents is higher if it is nearly invisible. Another reason is that the plant located within the mountain is better protected than it would be outside. The third reason is a rather technical one, the downpipes are shorter.

Caverns have further tunnels that serve other functions. At first, a drive tunnel is needed to reach the different level in the power station with regular vehicles if machine parts need serving. Another one is needed for ventilation for the tunnel or the power plant. A third type of tunnel leads the wires to the necessary destinations and conducts electricity. Thus, there are four different types of tunnels at total: the drive tunnel, the ventilation tunnel, the electricity tunnel, and, of course, the water-bearing pipe.

Fig. 4.9 Pressure pipes of a storage power plant in Black-Forest region, Germany (see also Fig. 4.5)

Figure 4.10 shows a cavern of a hydropower plant in Goldisthal, Germany. The use of cavern allows the power plant to be situated under the hill or mountain. Such a power plant is not visible from anywhere outside.

4.3 Auxiliary Parts

In addition to the main parts and structural parts described in the previous two sections, there are several parts that neither directly take part in power generation nor constitute any structural element of the power plant. However, their use is very important for operation and control of the hydro power plants. In the following the components are described that are required for a smooth and efficient operation of a hydropower plant:

Fig. 4.10 Entrance of a drive tunnel of a cavern power station from inside and outside

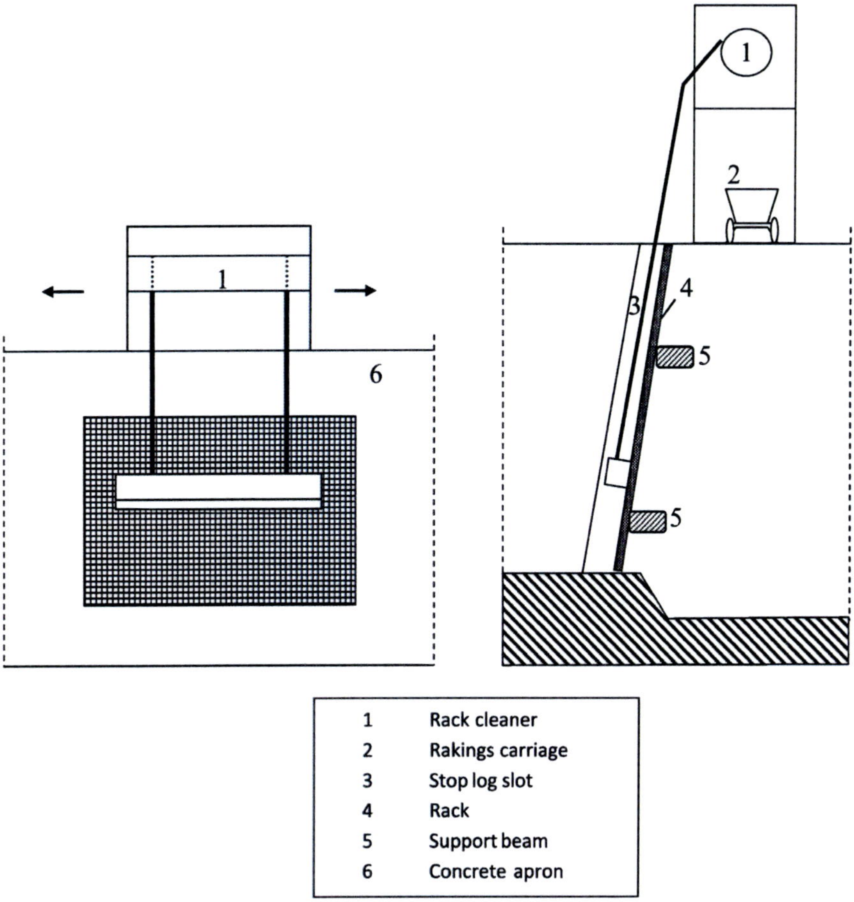

Fig. 4.11 Grill cleaning machine (concept of the figure taken by courtesy of König/Jehle [1])

4.3.1 Screening Grill

The first device faced by the stream of water moving towards the turbine is a grill (Fig. 4.11). This part has the function to protect living species or water life, which means that any fish or water animal as well as any solid such as a piece of wood or ice that are present in the water and are bigger than the holes in the grill will be prevented from entering into the power plant. If no such screening is present, the entry of such items into the turbine may cause damage to the blades. However, this grill causes a certain loss of energy through friction offered to the flowing water.

While the grill "collects" items that are larger than the gaps in the grill, such as branches or rubbish, it also causes additional resistance to the flow which is certainly not good for the performance of the power plant. In order to get rid of the collected items, the power plant has a grill cleaning machine. A shovel-shaped crossbar at the bottom of the grill is pulled up and fills its content into a wagon, which will then be disposed.

4.3.2 Control Gate

Control gates play a very important role by regulating the amount of water flow into the turbine through the penstock. These gates are normally of the vertical lifting type and due to their heavy weight and large size, can only be lifted with the help of large motors mounted on the top portion of the dam. The major part of the control gates remains submerged in the water body. As it is constant contact with corrosive conditions, the material of the control gates of a hydropower plant is very important and critical. For the same reason the maintenance and repair of the control gates is a work-filled job related to hydropower plants.

4.3.3 Control and Shut-Off Valves

Shut-off valves are needed in hydroelectric power plants for interrupting the water flow during operation. This can be necessary for safety issues concerning the bottom water, for draining the turbine if the turbine needs repair or service, as well as in a pumped storage power plant, when switching between turbine working and pump working.

There are three main types of shut-off valve explained below and shown in Fig. 4.12. The first valve to be mentioned is the ball valve. This is similar to a ball with a hole inside. When the water is supposed to flow, the hole of the ball is aligned with the direction of flow and water rushes through the hole. If it is required to stop the flow, the ball is turned by 90°.

Figure 4.13 shows a photograph of a control valve of a hydropower station. On the left side of the picture, the connection from the penstock is visible, and on the right water is entering the turbine spiral casing. The valve is very heavy in order to withstand the pressure of water, 30 bar in this case. It requires a special driving mechanism for its operation; cylinders and lever for controlling the position of the valve is also seen in this figure.

Another variant of shut-off valves is the throttle valve. This is a disc which remains parallel to the direction of flow when flow is to be allowed. It is vertically turned by 90° for stopping the flow. In order to withstand the strong force of the water while stopping the flow, the disc is shaped like a lens, thicker in the centre and gradually becoming thinner towards the edges. The position of the disc is changed by a mechanism through a long lever which is outside the pipe and is connected to the lens-shaped valve.

The third kind of shut-off valves is the turn valve which is similar to the throttle valve, the only difference being that it is turned from another position.

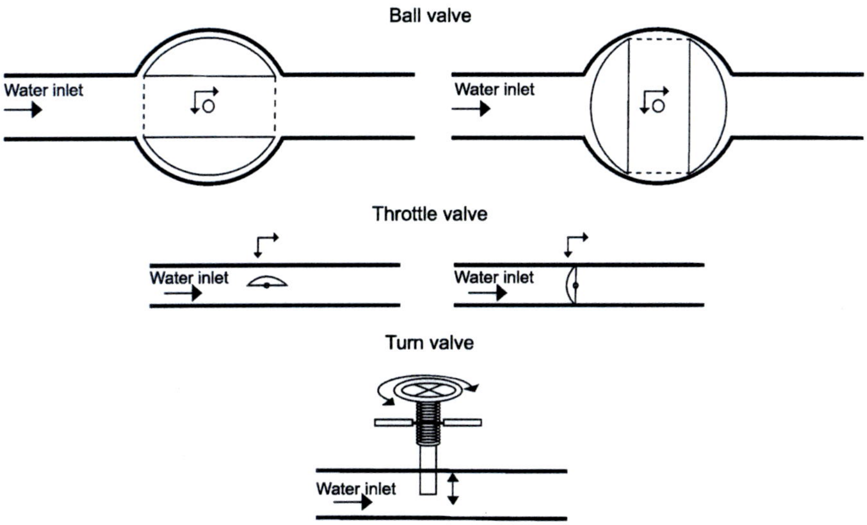

Fig. 4.12 Types of shut-off valves

The functions of the different valve types are identical; although the respective constructions vary. In large plants, moving the ball, valve or throttle valve is achieved with the help of electric motors and a hydraulic or pneumatic system. Turn valves are not commonly used in daily operation.

4.3.4 Fish Passes

When a dam is constructed for a river power plant, man interferes with nature and the natural habitats of various species. The creation of the dam not only widens the river to take on the shape of a water reservoir, but blocking the water with a dam also prevents water-living animals from going from one side of the dam to the other. In order to protect the life of water animals, and to enable water animals to pass river power plants from upstream to downstream and vice versa, fish passes are provided.

The principle of fish passes is shown in Fig. 4.14. A fish pass provides stair-like structure besides the dam. These are a series of small water chambers, one after the other, with a deflector located alternately on the left or right of the dam (Fig. 4.15). The alternate deflectors help to keep the velocity of the water stream relatively low so that fishes can survive while passing through them. Thus the fishes can enter either side of the dam by jumping from one chamber to the next.

In the case of more recent power stations small artificial brooksrunning beside the dam will are used as fish ways.

Fig. 4.13 Control valve of hydropower plant at Roenkhausen, Germany

The potential energy of water flowing through the fish passes is not utilized for power generation, hence it is considered as a loss of energy. However, this is a necessary contribution to the maintenance of the surrounding nature of a river power plant.

4.3.5 Guide Vanes

A ring of stationary adjustable blades or vanes (see Fig. 4.16) constitutes an important part for controlling the operation of a hydro-turbine. These vanes are called guide vanes since they guide the water in the most suitable direction with respect to the moving blades of the turbine so that the water enters at the desired angle. Guide vanes receive water from the spiral casing or trumpet inlet.

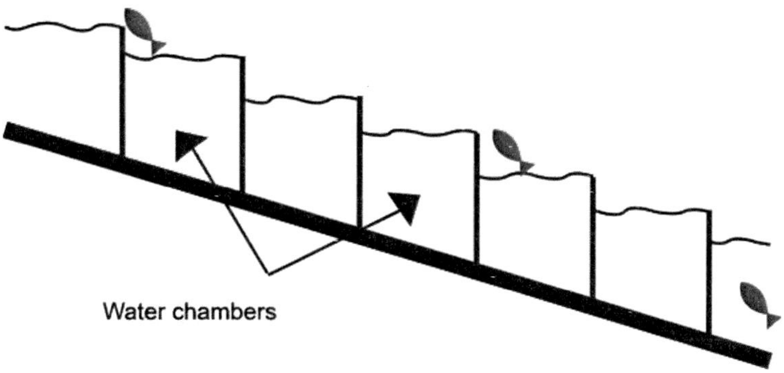

Fig. 4.14 Scheme of a fish pass

Fig. 4.15 Fish pass of a power station in Southern Germany (*winter time*) (photograph by courtesy of R. Lenk)

Right after the stationary vanes the water flows through the guide vanes which turn the water onto the turbine. The guide vanes are able to regulate the flow rate by leaving the space between each vane more open or more closed. The vanes are

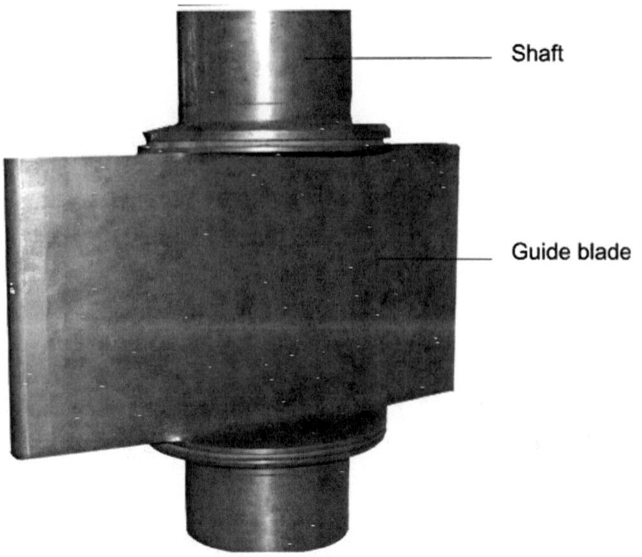

Fig. 4.16 Adjustable guide vane with shaft (hydropower plant Palmiet, South Africa)

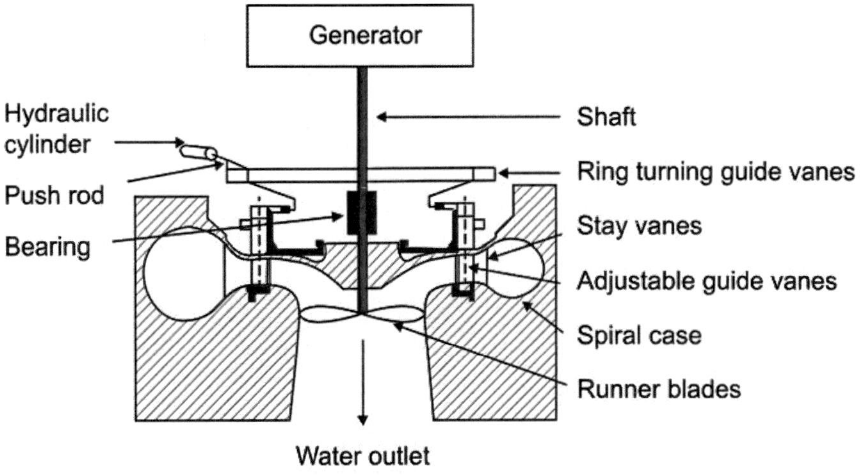

Fig. 4.17 Scheme of operating mechanism of guide vanes

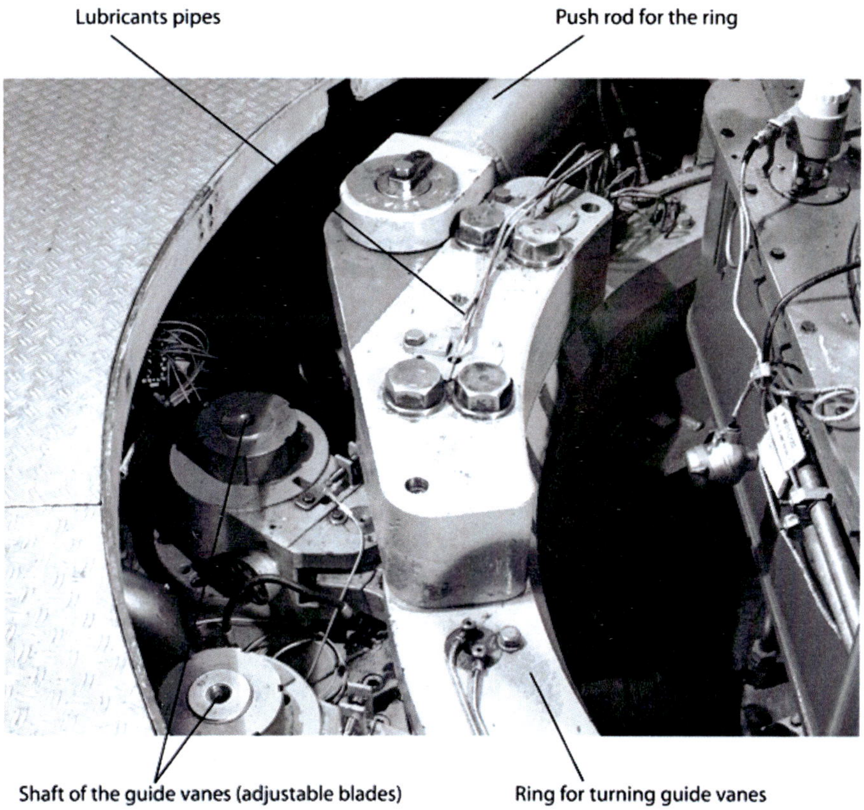

Fig. 4.18 Operating mechanism of guide vanes of power station at Roenkhausen Germany

moved hydraulically with the help of the adjusting ring which is turned in clockwise direction and anticlockwise direction by a central impetus.

After passing through the guide vanes water reaches the turbine blades, also called rotor vanes or runner vanes. Here the power of the streaming water drives the turbine by pushing the sloping shape of the vanes. As mentioned in sector describing the generator the number of rotations per minute has to be constant because of the constant grid frequency. The exact rotation of turbine and generator is controlled by the shape of the rotor blades as well as by controlling the flow rate of the water, which has been regulated by turning the guide vanes.

Figure 4.17 shows the schematic diagram of the operating mechanism of guide vanes. For operating the guide vanes, the big push rod (seen on the top of Fig. 4.18), turns the ring, on which all the vanes get swilled by the same angle around the individual shaft of every guide vane. It may be noted from the figure that the mechanism requires involvement of several sub-systems such as a lubrication system.

As mentioned at the beginning of this chapter, a hydropower plant may have thousands of minor parts, however, the parts explained in this chapter are the most important ones only, which are required to understand the working and hydro control of power stations.

Reference

1. von König/Jehle (2005) Bau von Wasserkraftanlagen—Praxisbezogene Planungsgrundlagen. In: Müller CF (ed) 4th edn. Heidelberg 2005, ISBN 3-7880-7765-4

Chapter 5
Hydraulic Turbines: Types and Operational Aspects

5.1 Classification of Hydraulic Turbines

The turbine is considered to be the heart of any hydropower plant since it converts the power of water into rotation of a shaft which, through a generator, is capable of producing electricity. Since the key lies in the efficient conversion of the power of water into rotation, the proper selection and operation of the turbine is very important.

Turbines of hydro power plants can be classified in many ways. Three major criteria for classification are:

- Classification based upon direction of flow
- Classification based upon pressure of water
- Classification based upon shape and orientation of turbine

They are described below.

5.1.1 Classification Based Upon Direction of Flow

Water can pass through the hydraulic turbines through different flow paths. Based upon the path of water flow, hydraulic turbines can be categorized into three types:

- Axial flow turbines
  In this category of hydraulic turbines the water flow path is mainly parallel to the axis of rotation. Kaplan and Propellor turbines are the most popular types of such turbines.
- Radial flow turbines
  This type of hydraulic turbines has water flowing mainly in a plane that is perpendicular to the axis of rotation of the blades. Such a type of turbine is the Pelton turbine.

H.-J. Wagner and J. Mathur, *Introduction to Hydro Energy Systems*,
Green Energy and Technology, DOI: 10.1007/978-3-642-20709-9_5,

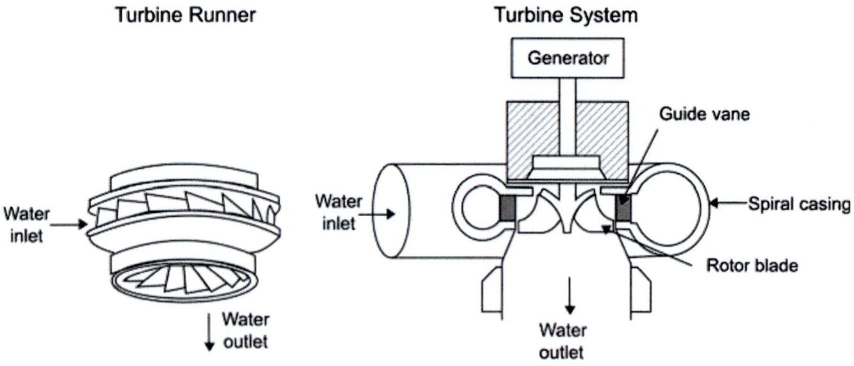

Fig. 5.1 Scheme of a Francis turbine

- Mixed flow turbines
  In practice, for most of the hydraulic turbines, the direction of flow is neither purely axial nor purely radial. It has a significant component of both axial and radial flows. Such types of hydraulic turbines are called mixed flow turbines. The Francis turbine (see Figs. 5.1 and 5.2) is the most popular type of mixed flow turbine in which water enters in radial direction and exits in axial direction.
- Crossflow turbines
  In this type, water runs through the blade ring of the cylindrical rotor (turbine wheel), which looks like a blower wheel of an electric air heater. Here, water gives energy twice, to the upper turbines blades and the lower turbine blades. A popular type of this turbine is a Banki turbine or an Ossberger turbine (see Fig. 5.3), called after their designers. A sliding valve controlled the amount of water feeding the turbine. Crossflow turbines are only used in the lower power range, i.e. below 1 MW. Where the water supply requires, the Ossberger-crossflow turbine is built as a multi-cell turbine. The normal division in this case is 1 to 2. The small cell utilises small and the big cell medium water flow. This explains why greatly fluctuating water supplies could realized by high efficiency.

5.1.2 Classification Based on Pressure Change of Water

Another method for classifying hydraulic turbines is based upon the change in the pressure of water when it passes through the rotor of the hydraulic turbines. Based on the pressure change hydraulic turbines can be classified by two types, the impulse turbines and the reaction turbines

5.1.2.1 Impulse Turbines

In this type of turbines, the pressure of liquid does not change while flowing through the rotor of the machine. In impulse turbines pressure change occurs only

Fig. 5.2 Installation of a Francis turbine runner at hydropower station *ITAIPU* (photograph by courtesy of Voith Hydro Holding GmbH & CO. KG)

in the nozzles of the set up that are not part of the rotor. In an impulse turbine, fluid is sent through a nozzle so that most of its potential energy can be converted into kinetic energy. The high speed jet then impinges on bucket shaped vanes mounted over a rotating shaft, converting kinetic energy of fluid into rotary movement of shaft. The most popular type of impulse turbine is the Pelton turbine (Fig. 5.4) or popularly known as Pelton wheel (Fig. 5.5).

5.1.2.2 Reaction Turbines

In this type of turbines, the pressure of water changes due to a change in the profile of the flow path while it passes along the rotor blades. The change in fluid velocity and reduction in its pressure causes a reaction on the turbine blades, which is the operating principle of these turbines. The reaction of fluid on blades rotation of the turbine is observed; hence they derive their name of reaction turbine. Francis turbines and Kaplan turbines are two popular types of reaction turbines.

The difference between the two types can be understood by assuming the shaft to be a hinged door. Impulse turbines present a situation where something strikes the door with high velocity, and as a result of the impact, the door opens. In the case of the reaction turbine the situation is similar to gently pushing the door, and

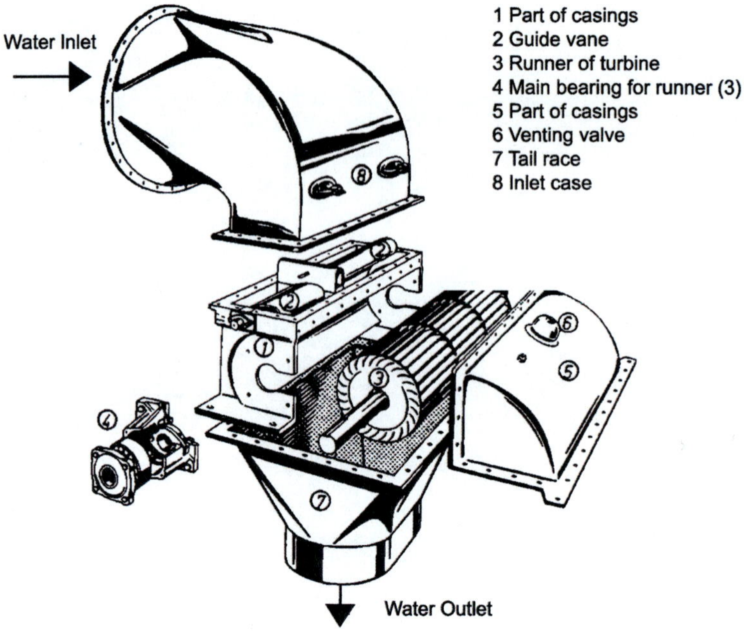

Fig. 5.3 Scheme of a Crossflow turbine (concept of the picture taken by courtesy of Ossberger GmbH & Co. KG, www.ossberger.de)

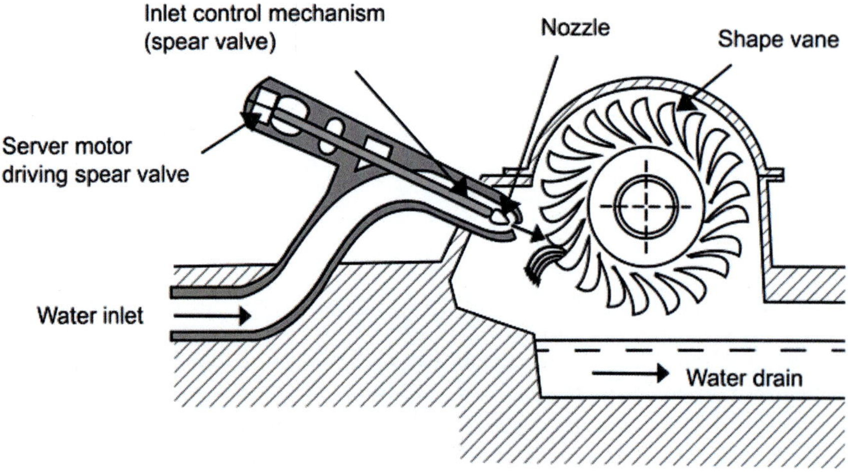

Fig. 5.4 Scheme of a Pelton turbine

changing the direction of the pushing force together with the changing orientation of the door at every intermediate stage. As in the first case, the higher the velocity of the striking object, the faster will the door open; similarly, with an impulse

Fig. 5.5 Pelton wheel (photograph by courtesy of Voith Hydro Holding GmbH & Co. KG)

turbine, the higher the velocity of water striking the bucket, the faster it rotates and, consequently, provides more power. In the other case, the force needs to be high and needs to act in the proper direction similarly to the case of the reaction turbine. More details about the two types are explained in separate sections.

5.1.3 Classification Based Upon Shape and Orientation of Turbines

Turbines can also be categorised by their construction or installation. There are bulb turbines, vertical turbines, and Straflo turbines. Bulb turbines are oriented nearly horizontally and their generator is located in a case shaped like a bulb or a pear. Here the streaming river water surrounds the construction partially and then flows through the turbine (see Fig. 5.6). This concerns Propeller and Kaplan turbines.

The Straflo turbine is an advanced bulb turbine, so to speak, with the generator poles on the outer ring of the rotor. The rotor blades are fixed to a ring that activates the generator. Straflo turbines are regarded as Propeller turbines.

Vertical turbines, on the other hand, are oriented nearly vertically and the generator is positioned above the water current (see Fig. 5.7). Examples of these types are found in Propeller turbines and Kaplan turbines.

The Kaplan and Francis turbines can either be positioned horizontally or vertically. Most of the big Francis turbines are found to be rotating in horizontal plane

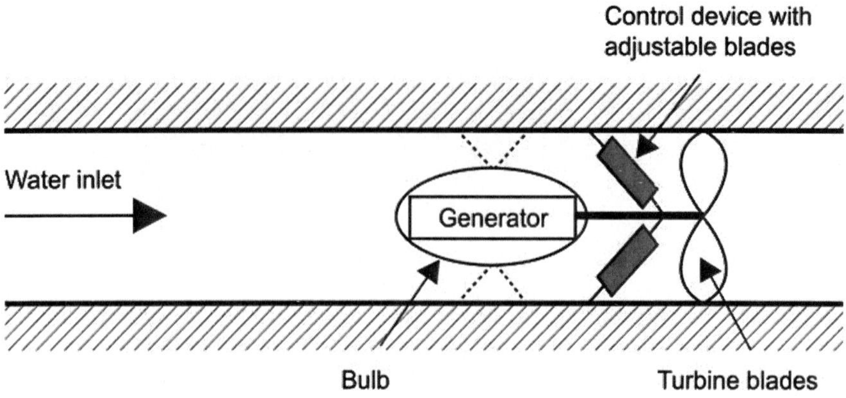

Fig. 5.6 Scheme of a bulb turbine (horizontally turbine)

due to better functioning of spiral casing in horizontal position. Pelton turbines are turbines used for high altitudes but low masses of water. These are mostly rotating in vertical plane.

In addition to the methods of classifying turbines described above, turbines can also be classified by the degree of loading. They can either be fully or partially loaded. Full loading means that the water streams to the complete rotor periphery, while partial loading occurs when the water streams only on a few areas to the rotor periphery. As a consequence, turbine blades with partial loading have a "rest" period between the loadings. An exemplary turbine of this type is the Pelton turbine, in which only a small number of blades are operational at any single point of time.

Another categorisation based on the pressure gradient upon the rotor. There are overpressure turbines, like in Kaplan, Francis or Propeller turbines. Here the pressure is a result of the accumulation of water on the turbine, and the pressure is higher on the upper water side than on the lower water side. The other type is the equal pressure turbine, where the water flows through the open air before reaching the turbine blades. Compared to the overpressure turbines, equal pressure turbines have a higher water pressure on the turbine blades than overpressure turbines, because there are bigger heights of fall.

5.2 Theory of Hydroturbines

In addition to the former section, turbines can also be classified by their type of construction, the most important ones being the Francis, Pelton and Kaplan or Propeller turbines, which will be described further in the following chapters. They are named after their inventors.

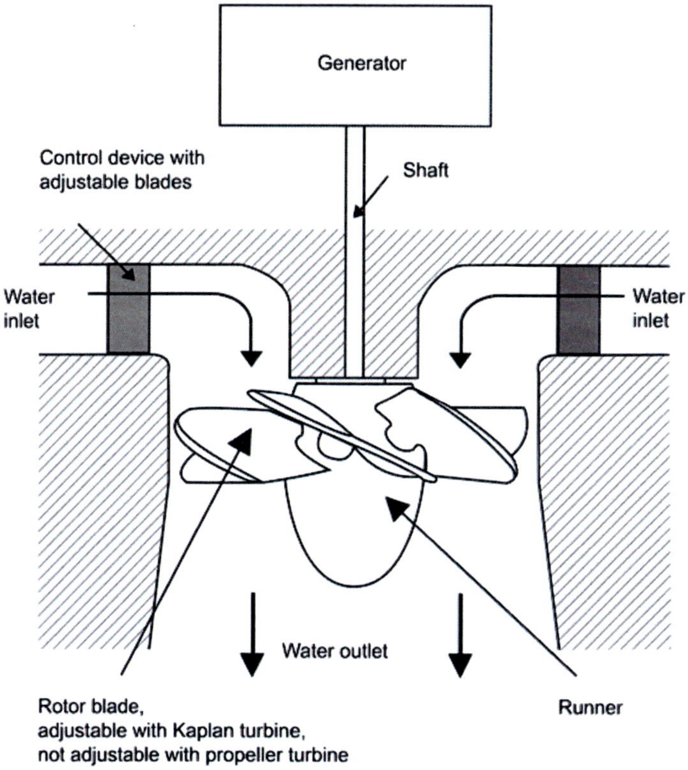

Fig. 5.7 Scheme of a vertical turbine

5.2.1 Francis Turbines

The Francis turbine was developed by James Francis in England in 1849. It resembles a ring cake form (see Figs. 5.8 and 5.9). Francis turbines are commonly used for middle heights of drop from about 15 up to 500 m and a water flow of up to about 500 m^3/s. In practice, Francis turbines are often used in pumped storage power plants, as they can also be used for pumping.

The Francis turbine is a reaction turbine. The water moves through the turbine, giving up its energy. The water flow changes its momentum and gives its energy to the surface of turbine blades that causes change of pressure due to their profile. The turbine is located between the high pressure water source and the low pressure water exit.

Fig. 5.8 Francis turbine runner (photograph by courtesy of Voith Hydro Holding GmbH & Co. KG)

Theory of Francis Turbine

Adjustable guide vanes cause the water to enter the runner at point 2 with a distinct angular momentum in the direction of rotation (see figure) with the velocity vector $\vec{c}_2$. The curved runner blades force the water to leave the runner at point 1 with velocity vector $\vec{c}_1$. The runner rotates at the peripheral speeds $\vec{u}_2$ and $\vec{u}_1$. From the perspective of the rotating runner, the fluid velocity vector is $\vec{w}_1$ and $\vec{w}_2$ (*see velocity diagrams*).

The best efficiency point is reached when the vectors $\vec{w}_1$ and $\vec{w}_2$ are parallel to the runner blades.

The change of the angular momentum of the fluid between runner inlet and runner outlet cause torque acting at the rotor blades.

The angular momentum change per unit mass of fluid on point 2 is described by the product of the radius r_2 and the component c_{u2} of velocity vector $\vec{c}_2$ in the direction of the peripheral speed u_2.

Fig. 5.9 Manufacturing of a runner of a Francis turbine in a Chinese factory (photograph by courtesy of Y. Ma)

Theoretically, the power P_F delivered from the fluid to the blades is the product of torque T_F and angular velocity ω. The torque T_F is expressed by the equation

$$T_F = \dot{m}(r_2 \cdot c_{u2} - r_1 \cdot c_{u1}) \tag{5.1}$$

where:

$\dot{m}$ mass flow of water
r_2 radius of inlet point
r_1 radius of outlet point
c_{u1} component of velocity vector $\vec{c}_1$ in direction of peripheral speed $\vec{u}_1$
c_{u2} component of velocity vector $\vec{c}_2$ in direction of peripheral speed $\vec{u}_2$

The mechanical power of a Francis turbine is, as mentioned before, expressed by the following equation:

$$P_F = T_F \cdot \omega = \dot{m}(u_2 \cdot c_{u2} - u_1 \cdot c_{u1}) \tag{5.2}$$

where
u_2, u_1 peripheral velocities of runner at water inlet point 2 and water outlet point 1

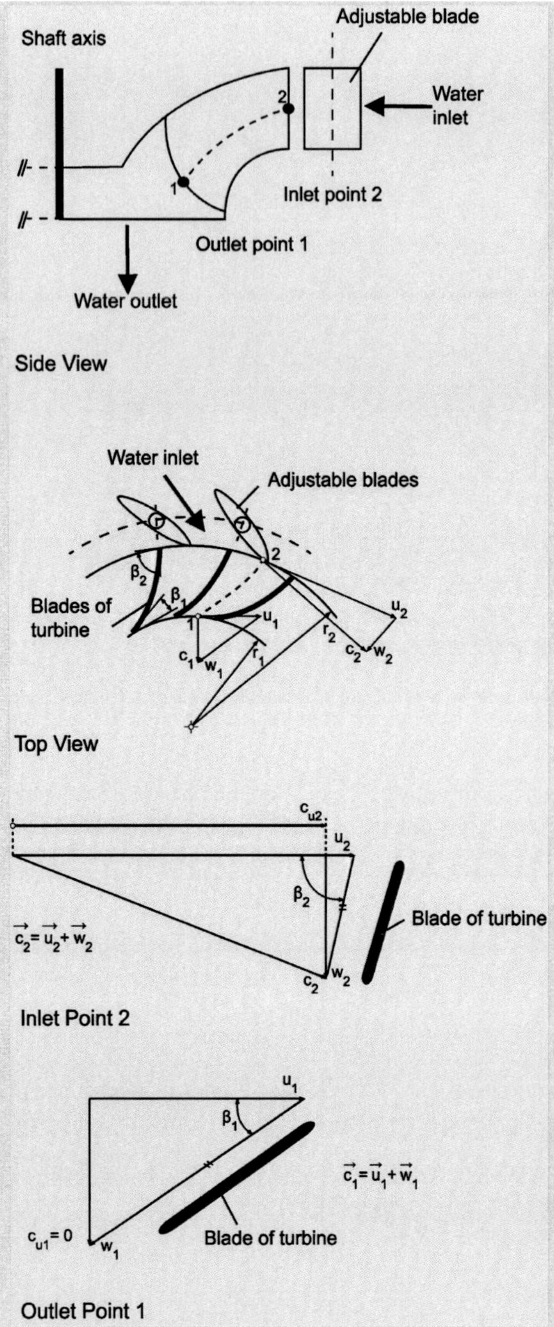

Velocity diagrams of Francis turbine [1]

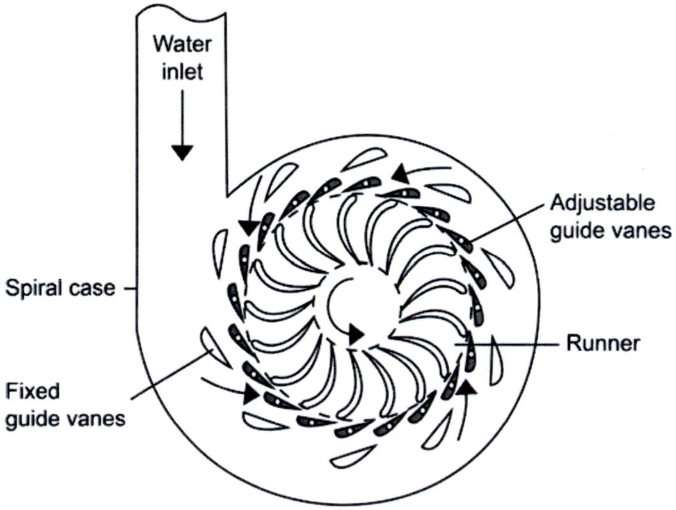

Fig. 5.10 The system of blades of a Francis turbine (*top view*)

The inlet of water into the turbine possesses a spiral shaped entry. The water looses part of its pressure in the spiral casing for maintain its speed. Guide vanes, situated immediately next to the runner (see Fig. 5.10), direct the water tangentially to the turbine wheel, without significant impact. This radial flow acts on the vanes of the runner, forcing it to rotate. The reason for rotation of the turbine is the pressure reduction of water along the runner blades which import a force on the blades. Water exits the turbine through the draft tube, which also acts as a diffuser and reduces the exit velocity of the water. This is done to recover a maximum of energy from the flowing water that helps improving the efficiency of the turbine. When water flows through the outlet it has made a 90° turn compared with the inlet. The guide vanes (or wicket gate) may be adjusted to allow an efficient turbine operation for a range of water flow conditions and for controlling the speed of the runner by changing the angle of the blades. Power is transferred from the runner blades to the shaft of the turbine in the form of torque and rotation. The shaft of turbine is directly connected to the generator for obtaining electric power.

The size and shape of the blades of a Francis runner depend upon its specific speed (see box). Higher specific speed means lower rotation and a bigger construction height of the Francis runner (see Fig. 5.11). This means that for a given power output, the runner should admit a comparatively large quantity of water and at the same time the velocity of discharge at the runner outlet should be small to avoid cavitation.

Due to a large discharge area (area perpendicular to the axial direction), this type of runner can pass a large amount of water through itself meeting a low exit velocity.

Fig. 5.11 Runners of two Francis turbines (laboratory size in kW-power range). The one on the *left* is designed with low specific speed and the one on the *right* with high specific speed

Specific Speed

Originally specific speed n_s is a non-dimensional number, often used to classify pump impellers as to their type and proportions. The number is also used to classify hydraulic turbines. There are different definitions and interpretations of specific speed, also depending on the unit being followed in calculations. This number may loosely be expressed as a "Speed" only because the performance of the reference device is linearly dependent on its speed. The specific speed is the rotation of the reference device.

As a reference device—in metric units—a small turbine is chosen which is designed for a water volume V of 1 m^3/s and a water fall height h (hydraulic height) of 1 m in metric units.

The rotation of real turbine n can be calculated with the empirical relation

$$n = n_s \times \frac{h^{3/4}}{Q^{1/2}} \tag{5.3}$$

where

n, n_s in 1/min
h in m
Q in m^3/s

Fig. 5.12 Installation of Pelton wheels at storage power station Walchensee, Germany (photograph by courtesy of Voith Hydro Holding GmbH & Co. KG)

5.2.2 Pelton Turbines

The Pelton turbine (Fig. 5.12) was developed by Lester Pelton in the USA in the year 1889. This type of turbine is useful for high heights of water drop in the range of 200–2,000 m and a low volume flow of up to about 40 m^3/s. A typical Pelton turbine resembles water wheels that were used in former times.

The Pelton Turbine has a circular disk mounted on the rotating shaft or rotor. This circular disk carries cup/bucket shaped blades, called buckets, placed at equal spacing at its circumference (see Fig. 5.13). Single or multiple nozzles are arranged around the wheel such that the water jet emerging from a nozzle moves in a direction tangential to the circumference of the wheel of the Pelton turbine so as to strike the buckets (see also Fig. 5.4). According to the available water head

Fig. 5.13 Shaped blades with splitter of a Pelton wheel

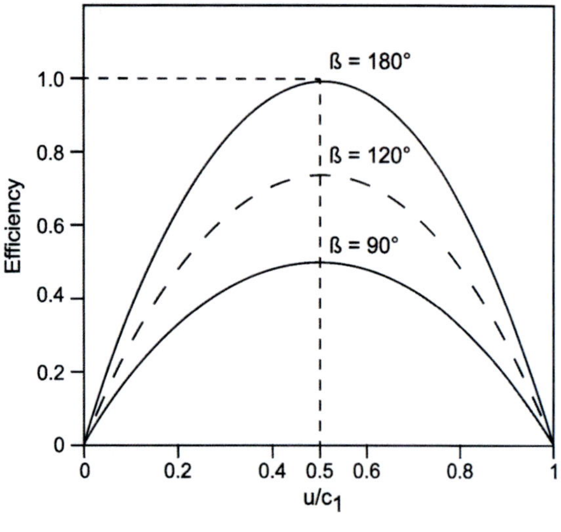

Fig. 5.14 Efficiency of a Pelton runner depending on velocities and shape of bucklets

(pressure of water) and the operating requirements the shape and number of nozzles are placed around the Pelton wheel.

In a Pelton turbine, also known as Pelton wheel, water jets impact on the buckets (also called blades) of the turbine making the wheel rotate, producing torque and power. The entire assembly that rotates due to water striking the blades is called runner.

Theory of Pelton Turbine

The high speed water jets emerging from the nozzles strike the buckets at splitters, placed at the middle of a bucket, with a velocity $\vec{c}_1$ from where the jets are divided into two equal streams. These streams flow along the inner curve of the bucket and leave it in a direction nearly opposite to that of the incoming jet with velocities $\vec{w}_2$ and $\vec{w}_3$ which ideally have the same scalar value of velocity as that of the incoming water steam. The buckets are moving with the peripheral speed u, they were hit by water with the speed $\vec{w}_1$ (see Figure).

Velocities of water at various points are explained by the so called velocity diagrams (see Figure).

The change in momentum due to the loss in velocity and change in direction of the water stream produces an impulse on the blades of the wheel of the Pelton turbine. This impulse generates the torque and causes rotation in the shaft of the Pelton turbine. To obtain the optimum output from the Pelton turbine, the impulse received by the blades should be maximized which required maximizing the change in momentum of the water stream before it leaves the buckets or blades.

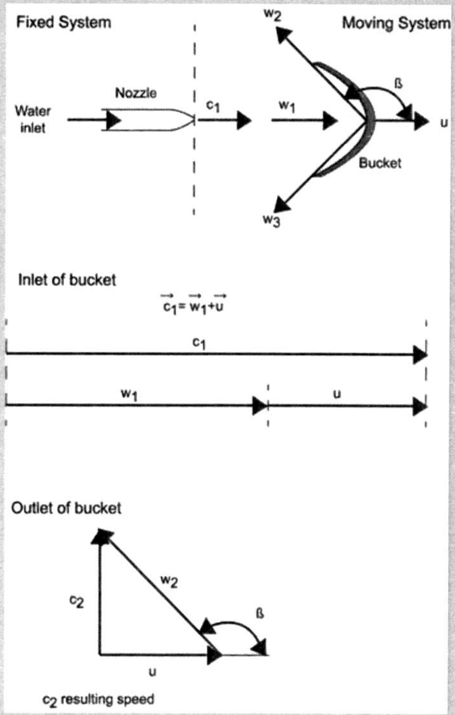

Velocity diagrams of Pelton turbine.

According to theory, the efficiency of a Pelton turbine depends on the angle β, water inlet velocity $\vec{c}_1$ and peripheral speed $\vec{u}$. Figure 5.14 shows the maximum efficiency by

$$\beta = 180^\circ$$

and

$$\frac{u}{c_1} = 0.5$$

The mechanical power of a Pelton turbine wheel P follows the equation

$$P = u \cdot \dot{m} \cdot (c_1 - u) \cdot (1 - \cos \beta) \tag{5.4}$$

where

c_1 velocity of water from nozzle
u peripheral speed of Pelton wheel
β angle of buckets (see Velocity diagram page 85)
$\dot{m}$ mass flow of water

In actual conditions, apart from the profile losses, other losses such as mechanical friction, aerodynamic drag, nozzle losses, also occur and restrict the overall efficiency of Pelton turbines to around 90%.

A typical setup of a system generating electricity by using Pelton turbine will have a water reservoir situated at a significant height above the Pelton wheel. The water from the reservoir flows through a pressure channel to the penstock head and then through the penstock or the supply pipeline it reaches the nozzles through which the water comes out in the form of high speed jets striking the blades of the Pelton turbine.

For a constant water flow rate from the nozzles the speed of the turbine changes with changing loads on it. For a good frequency quality of hydroelectricity generation—that means constant frequency-, the turbine is required to rotate at a constant speed. To keep the speed constant despite the changing loads on the turbine, the water flow rate through the nozzles is changed. To control the gradual changes in load, servo controlled spear valves are used in the jets to change the flow rate. For a sudden reduction in load the jets are deflected using deflector plates so that some of the water from the jets does not strike the blades. This prevents over-speeding of the turbine.

Since water and most liquids are nearly incompressible, most of the available energy is extracted in the first stage of the hydraulic turbine. Therefore, Pelton wheels have only one turbine stage, unlike gas turbines that operate with compressible fluid.

Pelton wheels are the preferred turbine for hydro-power, when the available water source has relatively high hydraulic head at low flow rates. Pelton wheels are made in all sizes. The largest units can be up to 500 MW. In addition to the big ones there are also small Pelton wheels with a diameter size of less than half a meter and the power of only a few kilowatts, and can be used to tap power from mountain streams having flows of a few m^3 per minute.

Fig. 5.15 Kaplan turbine runner (photograph by courtesy of Voith Hydro Holding GmbH & Co. KG)

5.2.3 Kaplan Turbine and Propeller Turbine

The Kaplan turbine was developed by Viktor Kaplan in Austria in 1912. The rotor looks like an ordinary propeller of a ship with movable rotor blades (see Fig. 5.15). Its major special features are low heights of drop till about 25 m, even if it could work with a higher drop height, and adjustable rotor blades for changes in volume from about 1 m^3/s up to 500 m^3/s.

The Propeller turbine, on the other hand, is similar to the Kaplan turbine in terms of required height of water drop and volume flow as well as appearance, but it also shows differences. While it is used for a more or less fixed amount of water, its rotor blades are not adjustable as in case of Kaplan turbine. However, in huge power plants there is often more than one type of turbine in order to convert nearly all of the potential energy by switching on and off different numbers of Propeller turbines at any time.

Except for the direction of water flow, both Kaplan and Propeller turbines, also fall into the category of reaction turbine. Their working principle therefore, also is quite similar to the working of the Francis turbine.

Kaplan turbines and especially Propeller turbines are often found in river power plants, as they are the best of all turbine types to deal with a high water flow with low height of fall.

To generate a substantial amount of power from small heights of water using the Kaplan turbine it is necessary to have large flow rates through the turbine. The Kaplan turbine is designed to accommodate the required large flow rates. Except for the alignment of the blades the construction of the Kaplan turbine is very much similar to that of the Francis turbine. The Kaplan turbine also has a ring of guide vanes at the water inlet to the turbine.

Unlike the Francis turbine which has guide vanes directly at the periphery of the turbine rotor, there could be a passage between the guide vanes and the rotor of the Kaplan turbine (see also Fig. 5.7). The shape of the passage is such that the flow which enters the passage in the radial direction is forced to flow in axial direction. The rotor blades are attached to the central shaft of the turbine. The blades are connected to the shaft with moveable joints such that the blades can be swivelled according to the flow rate and available fall height of water.

The inlet of water is in a shape of a scroll-shaped tube that wraps around the turbine's wicket gate. Water is directed tangentially through the wicket gate and spirals on to a propeller shaped runner, causing it to spin.

The outlet is a specially diverging draft tube that helps decelerate the water and recover kinetic energy.

Turbines could be used as mentioned before in both orientations, horizontally or vertically.

Variable geometry of the adjustable gate vanes and turbine blades allow an efficient operation for a wide range of flow conditions. Kaplan turbine efficiencies are typically about 90% and more, but may be lower in very low water height applications.

Propeller turbines have non-adjustable propeller vanes. They are used in conditions where the volatility of the water volume is not large. Commercial lay available Propeller turbines range from few kilowatts up to more than 100 MW.

Pit turbines are the Kaplan or Propeller turbines which are installed as bulb turbines. They have a gear box. These turbines usually have a very small capacity, or are in the order of micro or small power plants due to the presence of a gear box. Gear boxes restrict the maximum capacity to about 10 MW. However the most common design in the MW range and more is to couple the turbine directly with a generator with a multipole generator having fixed number of poles.

Straflo turbines are axial Propeller turbines in a very compact design. The tips of the blades are connected with a ring, on which the poles of the armature of the generator are fixed. Turbine runner and generator armature are one part. The generator stator is outside the water channel, connected to the periphery of the runner.

A critical element of Straflo turbines are the sealing between armature and stator.

S-turbines eliminate the need for a bulb housing by placing the generator outside of the water channel. This is accomplished with a 90° angle transmission in the water channel and a shaft connecting the transmission and generator.

Tyson turbine, also often called River turbines, are fixed propeller turbine designed to be immersed in a fast flowing river, either permanently anchored in the river bed, or attached to a boat or barge. There are only few installations of this

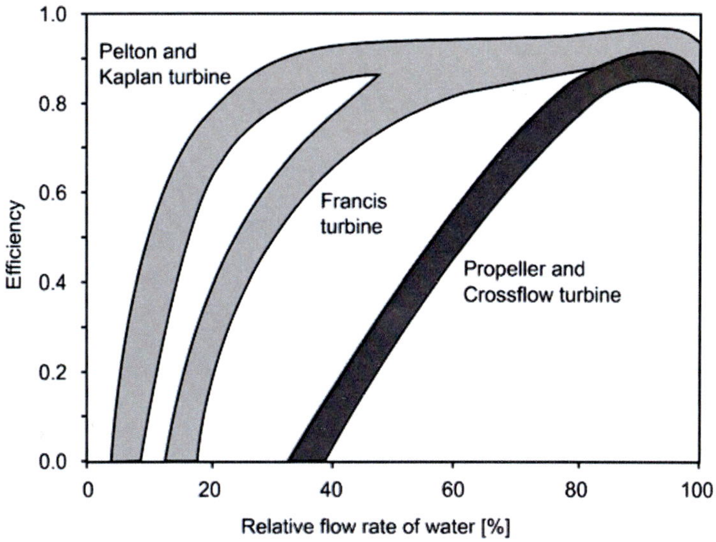

Fig. 5.16 Typical efficiency curves for different turbine types (the shape could change according to turbine design)

type worldwide since they are not suitable for rivers having a river depth flow of water below about 2 m/s.

5.3 Operational Aspects of Turbines

5.3.1 Efficiency

The efficiency of a turbine depends on the water flow and the type of turbine (Fig. 5.16). The efficiency of Pelton and Kaplan turbines is high over a wide range of water flow. Propeller and Crossflow turbines present a distinct optimum.

The following features are to be noted:

- The maximum efficiency of all the turbines (expect very small ones) is of the order of 90%, however, this maximum efficiency is not at 100% flow.
- The efficiency of all the turbines is low if the flow is very reduced. There must be a minimum of water flow for turbine operation (e.g. >30% of rated water flow for Francis).
- In the case of fixed propeller turbines, the efficiency drops very fast with reduction in flow.
- The shape of the efficiency curve is also dependent on the specific speed of a turbine.

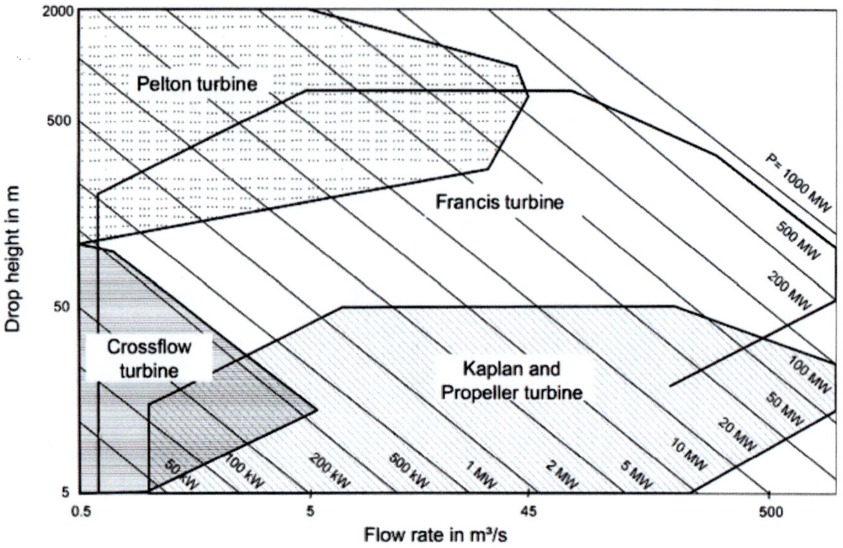

Fig. 5.17 Operation areas of hydro turbines and their power (logarithmic scales)

5.3.2 Selecting a Type of Turbine

If the head and flow of water are known, Fig. 5.17 can be used to develop an idea of which turbine is suitable for that particular combination.

The figure shows that:

- The Pelton turbine is suitable for low flow and high fall height.
- For medium to high fall height and medium to high flow, the Francis turbine is suitable.
- For low or medium flow and low or medium fall height, the Kaplan turbine is appropriate.
- There is an overlap between Pelton and Francis, and Kaplan and Francis turbines. This means, both types of turbines are suitable for such combinations of fall height and flow. However, the final decision must be based on all details of the real situation including especially the cost aspects.

5.3.3 Two-Block and Three-Block-Systems

The main machinery of a pumped storage power plant consists either of two or of three blocks. While the three-block-system has pump, turbine and generator as individual blocks, pump and turbine are combined together with a Francis Turbine

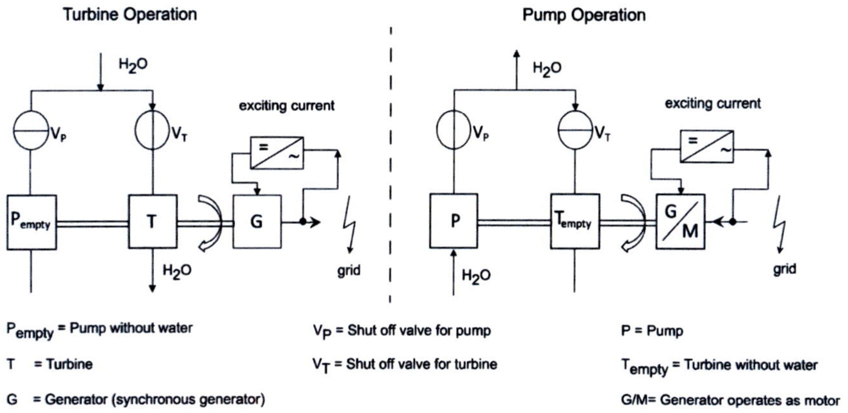

Fig. 5.18 Turbine and pump operation of a three-block-system

in the two-block-system. The advantage of the two-block-system is that the investment gets reduced since no separate pump is to be purchased. The disadvantage, on the other hand, is that the switching a turbine between pumping and power generating modes is more complicated than with the operation of a three-block-system. Another disadvantage of two-block-system is that selection of the Francis turbine, working as turbine as well as a pump may be a compromise in respect of the efficiency, especially if the flow and head combination suggests some other type of turbine for maximum efficiency.

The three-block-system has, as mentioned, three major machine parts (see Fig. 5.18). Further, it has two valves, one in front of the turbine and one in front of the pump. During the turbine operation, the valve in front of the pump is shut and the valve in front of the turbine is kept open. The water runs through the turbine, the turbine turns, and the shaft of the turning turbine runs the generator. The generator, finally, produces electricity which is given to the grid. Switching the machinery from turbine operation to pumping is easy. When the operator wants to pump up the water when the demand of electricity goes down or the when price of electricity becomes cheaper, the valve in front of the turbine is shut and subsequently, the turbine is emptied by blowing compressed air. If some water is left in the turbine, it acts as a brake and also this water gets evaporated during operation as pump which damages the turbine blades. After the compressed air has done its work inside the pump, it is taken away and water from downward side enters the pump by opening the valve in front of it. When the water pressure at the outlet of pump increases beyond the water pressure required to pump the water back to the storage height, the valve in front of the pump is opened. Then the lower level water is pumped up to the upper level. The common shaft between the turbine and the pump turns into the same direction, which means that the direction is not reversed although the operation of plant gets reversed.

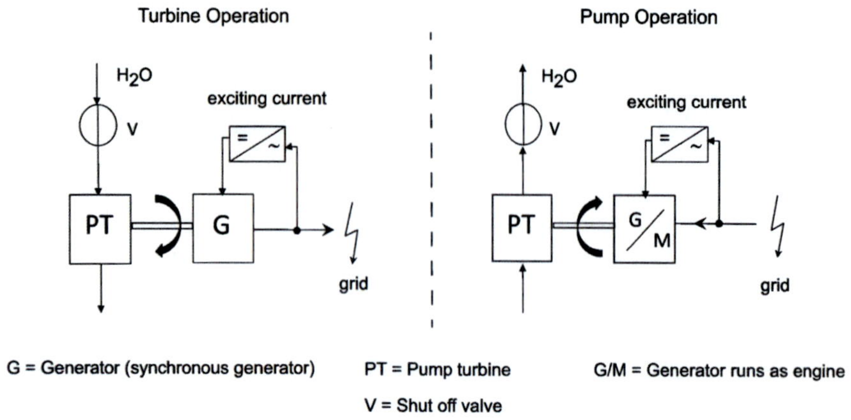

Fig. 5.19 Turbine and pump operation of a two-block-system

In the pumping operation, also called motor operation, the generator works as a motor when it takes electricity from the grid and turns the shaft for pumping. The turbine, which is also connected to the shaft, runs along empty.

When the mode is switched again, from the pumping operation to the turbine operation, also called generator operation, the operation or closing of the valves is done the other way round. At first, the valve in front of the pump has to be shut. Following it, the water inside the pumped is removed and sent back to the lower reservoir using the compressed air. As soon as the pump is empty, the other valve in front of the turbine is ready to be opened. After opening of the valve, the water first pushes the air inside the turbine out, and then the water flows through the turbine and the generator produces electricity in a regular way. In those modes, while the turbine turns the shaft, the pump runs along empty as well. Although the three-block-system has two valves, it never occurs that the turbine and the pump carry out their operation at the same time.

The operating logic of the two-block-system is somewhat different. As mentioned above, one single unit works as pump and turbine both (see Fig. 5.19). The Francis runner serves as turbine when the water comes from the upper level, and it works as a pump when water flow turns into the other direction. That is why only one valve is needed in this type of configuration. The switching from pumping to turbine operation works a bit differently as compared to the three-block-system. While switching the operation, the valve is first shut-off and the pumping is stopped. The valve has to be shut-off until the pump has completely stopped turning, as it has to turn into the other direction during the turbine operation. Then the valve is opened again and the pump now starts working as turbine, turned from the upper water generating electricity. For switching back to the pumping mode, the valve has to be shut again until the runner stands stop rotating. Then the runner starts moving in the opposite direction driven from the generator working as motor. When the water pressure inside the pump is higher than the static pressure at the valve from upper level water, the valve is opened. In case of a two-block-system no compressed air is required for emptying the turbine or pump.

In two-block-system, every switching means that the whole block system has to be driven down and up again. In earlier times, this used to need more time than in case of a three-block-system. Modern power stations with two-block-systems can change their operation mode in about 2–3 min.

As already mentioned in previous chapters, pumped storage power plants also have the task to deliver reactive power with the help of the synchronous generator.

For this operation mode, there is no difference between two- and three-block-systems. In this case all valves are completely shut and turbine and pump are running empty of water. Only the generator is exited in such a way that it delivers required reactive power to the grid. Energy losses by turning the pump and turbine are covered by active power taken from grid by generator.

Reference

1. Braitsch W, Haas H (1996) Turbines for hydroelectric power. In: Landolt-Börnstein (ed), Group VIII, vol 3. Energy technologies, Subvolume C, Chap 2.7, Renewable energy. Springer, Berlin, ISSN 1619-4802

Chapter 6
Use of Ocean Energies

6.1 Overlook

Previous chapters of this book have explained the technology for harnessing the power of water on land. In addition to this, there exists the opportunity for power generation in the water of oceans. This chapter will cover the technology involved in converting and utilizing the power of ocean. The various technologies for this purpose can be classified in five major categories:

- Tidal power plants
- Ocean current power plants
- Wave power plants
- Ocean thermal power plants
- Osmotic power plants

Except for tidal power plants, all listed technologies are in the research and development status of prototypes.

6.2 Tidal Power Plants

One of the most powerful forms of natural energy in the world is generated by the gravitation of the moon and the sun. This movement with respect to the earth causes low and high tides that happen twice within a 25 h period. The movement of the rising and falling sea level alters the potential energy of water that can be converted into electricity by the operation of a power plant. In order to use the potential energy a dam wall is created to enclose a certain amount of sea water in an artificial bay serving the purpose of reservoir of a hydro power plant. When the tide rises, the water enters the reservoir through the turbine which produces electric energy until the seawater inside the reservoir is almost as high as the

H.-J. Wagner and J. Mathur, *Introduction to Hydro Energy Systems*, Green Energy and Technology, DOI: 10.1007/978-3-642-20709-9_6,

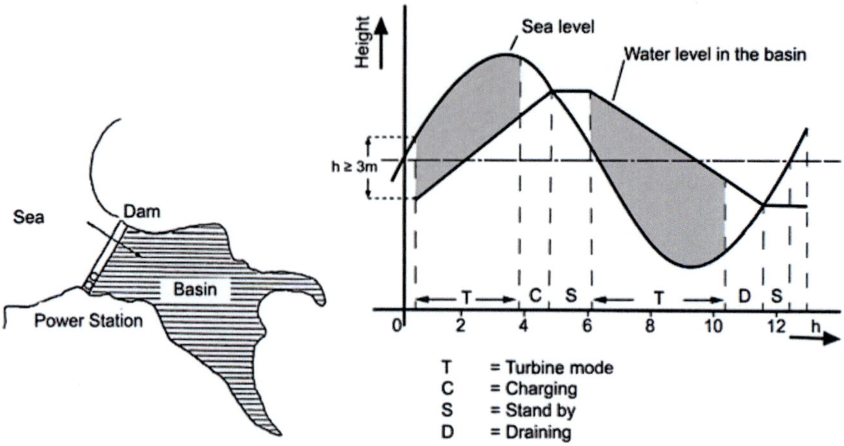

Fig. 6.1 Operation diagram of a tidal power plant

outside water level. At low tide outside the reservoir, the reverse process begins and the water inside the reservoir exits into the sea through the turbine. In practice there is a pause of about two hours between these two processes, because a minimum of water levels must exist to cover the energy losses inside the turbine. As the pressure of the water flow varies with the state of saturation of the reservoir, the power generation capacity of the flowing water is not constant (see Fig. 6.1). As a result of such variation, the disadvantage of using this type of energy is that it cannot be used permanently to cover a constant load because of the interruption. On the other hand, since the time and amount of water flow can be estimated accurately, the predictability of the power generation is an advantage of such systems.

The amount of electric energy that can be converted from the tides depends upon the amplitude of the sea level. For an efficient and economic use, the tidal energy needs to show an altitude of at least three metres from its average level. This means that regions with huge tidal ranges are especially suitable for an efficient and economic use of tidal power.

As a consequence axial flow turbines like Kaplan turbine are often used for tidal power stations.

6.2.1 Formation of Tides

Tides are created because the earth and the moon are attracted to each other due to their own gravitational pulls. The moon exerts a pull on everything on the earth to bring it closer and does the earth the same. The earth is able to hold onto everything from the pull of the moon, except for water. Since the water on earth is always moving, the earth cannot hold it firmly, and as a result the moon is able to pull it to some extent. Each day, there are two high tides and two low tides. The level of the ocean is constantly moving between high tide to low tide, and then back to high tide. There is about 12 h and 25 min between two consecutive high tides.

Since the position of the earth and the moon with respect to the sun changes throughout the year, the tidal power plant also produces a variable amount of energy as spring tide and neap tide differ in the regular sea level and supply different amounts of potential energy. When the moon is full or new, the gravitational pull of the moon and sun are combined. With this combination, the high tides are very high and the low tides are very low. Such situations are known as spring tides. However, they do not have anything to do with the spring season as one might infer from the name. They occur when the earth, the sun, and the moon are in a line. The gravitational forces of the moon and the sun both contribute to the tides. Spring tides occur during the full moon and the new moon.

The other tidal situation arises during neap tides. During the moon's quarter phases the pulls of sun and moon work at right angles, thus cancelling the effects of each other. This result in a smaller net pull acting on the water and consequently in a smaller difference between high and low tides which are very weak tides.

6.2.2 Existing Tidal Power Plants

Only two commercial scale tidal power plants are presently operational worldwide. One of them is in La Rance, France and the other one in Annapolis Royal, Canada. The tidal power plant located in La Rance was constructed in 1966 and is the first and still the biggest tidal power plant in the world. This location offers a difference in tidal altitude in the mouth area of up to 8 m quite regularly. This plant has 24 hydro turbines with an overall capacity of 240 MW that produce an annual electricity of 600 GWh.

The second important tidal power plant was constructed in 1984 in Canada at the southeast coast town of Annapolis Royal. The plant has single directional turbines that have a capacity to deliver 20 MW power. For creating a reservoir, the Annapolis River is blocked by a rock-filled dam. Though the location has an estimated potential to generate up to 5,000 MW power, the plans were restricted due to the colossal impact that the construction and the increase of the reservoir would have had on the local and regional environment.

Other tidal power plants run with a much smaller capacity. There are approximately 100 spots in the world where tidal power plants could be constructed to work efficiently.

The largest tidal power plant is being constructed in Sihwa in South Korea, south of Seoul. The power plant will contain ten turbines with a capacity of 26 MW each, resulting in the total capacity of the plant to reach 260 MW. An even bigger tidal power plant was proposed in Great Britain, at the end of Britain's longest river Severn that flows into the Atlantic Ocean between England and Wales. The plan would have included creation of a 16 km long dam between Cardiff and Bristol. But in 2010, Great Britain terminated the project claiming it would be too costly.

It is true that the biggest disadvantage of a tidal power plant is the high initial investment requirement for the power plant as compared to other power generation options. Another negative aspect of tidal power plants concerns the environment, both in a natural and an aesthetic sense. A huge reservoir and the construction of a concrete dam have a high impact on natural life on land and in the water. Nevertheless, the safe and calculable supply of electricity without any emission and independent of fossil fuels is the main motivation for constructing tidal power plants.

6.3 Ocean Current Power Plants

The term ocean current means the mass flow of the ocean water. When this mass movement, through its kinetic energy is used for power generation, it provides for ocean current power plants. These current power plants may be confused with tidal power plants due to a similar working principle, but there is a significant difference between the two. The mass movement of water can be either a short regional movement that is periodic (like the tides), or it can be a continuous movement that is not confined to a small region but is global, like the gulf. The reasons for these movements are quite diverse. One major reason is the friction of wind moving over the free water surface that forces the top by sun layers of water to move. Another reason for the movement of water is the so-called upwelling phenomenon, in which the upper warmer and nutrient-poor layers are exchanged with the lower cooler and nutrient-rich layers which causes a certain vertical current between the different layers, while the movement of the wind primarily causes horizontal streams.

Another reason for the movement of ocean water is due to the different warming up of the ocean water by sun. This effect gives rise to a vertical temperature gradient resulting in predominantly vertical streams. In addition, the density and salinity of water cause motion as they vary locally and mix with the surrounding water that does not have in these qualities. The available water streams are largely governed by the topography of the ocean, i.e., higher and deeper bottoms produces horizontal and vertical currents of water. Other forces that affect the current are collision with landmasses, the centrifugal force by the rotation of the earth, friction, and temperature.

For converting the current of water for the production of electricity, there is a wide range of technological approaches. The Italian ocean current power plant named *Kobold* (Fig. 6.2) was the first commercial oriented scale ocean current power plants. It was built in 2001 and is located in the Mediterranean Sea near, Italy. This tripod machine has a floating part that contains the generator and other

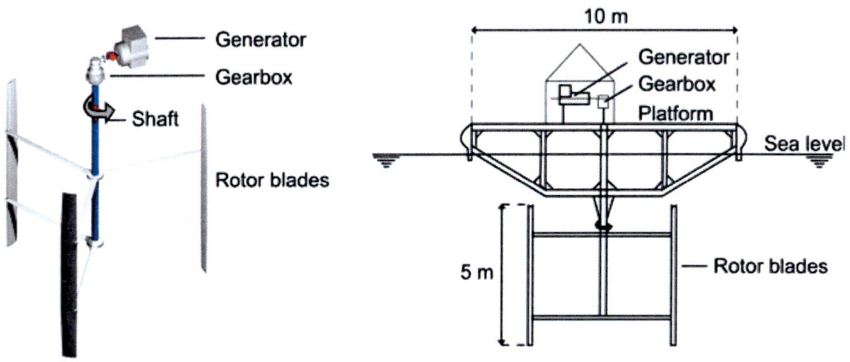

Fig. 6.2 Scheme of Kobold experimental ocean current power plant

equipment. It has several underwater parts that consist of a shaft, three propellers, and a frame for this underwater construction. The three propellers are quite similar to the rotor of a vertical axis wind turbine. In comparison to wind energy rotors the rotor dimensions of ocean current power plants can be smaller by same power because the density of water is much higher than the density of air. Due to the shape of rotor the power plant can utilise streams of water flowing in any direction. The rotor of Kobold plant has a diameter of six metres, and it requires a minimum flow velocity of 2 m/s to start turning. It has a maximum power generation capacity of 130 kW.

Stingray is another innovative experimental current power plant that is located on the sea bottom near Yell Sound off the Shetland Islands in the sea near Scotland. This construction has a four-legged frame with a holder for two wings which move with the current that uses vertical streams. The wings start moving from a speed of 2 m/s similar to the *Kobold* power plant, and its capacity is also 150 kW.

Modification of the previous concepts for ocean current power plants was demonstrated in the year 2003 when the British *Seaflow* was built as a test rig near the village of Lynmouth in the Bristol Channel by the British company Marine Current Turbines Ltd. This construction is similar to a horizontal axis wind energy converter as it has a rotor with two blades. The whole power plant is an underwater construction that converts horizontal water current into electricity. Rotor and generator are attached to a lifting collar, and an upper part is above the sea level. The test rig was operated until 2009, before a more commercial technology Seagen was introduced (Fig. 6.3). That is a similar model, also created by Marine Current Turbines Ltd. This power plant has a holder with two rotors. It was installed on the coast of Strangford Lough, Northern Ireland, in 2008, with a capacity of 1.2 MW.

This type of plant offers two advantages. First ocean energy can be used more efficiently by positioning the rotor in the layer with the most powerful current. Second, the collar can be lifted up above the sea level which makes repair and service related work easier. The disadvantage is the high cost of investment due to the construction. The *Seaflow* requires a minimum flow velocity of 2 m/s for starting to generate power and its rated capacity is 350 kW.

Fig. 6.3 Seagen ocean current power plant with graphic *inset* to show operation (photograph by courtesy of Marine Current Turbines Ltd.)

The minimum speed of water for *Seagen* should be 2–3 m/s. There is a wide range of spots in the oceans that offer current energy, hence thousands of kilometres of coasts could be used to build up such power plants. However, when planning the construction of ocean current power plants, deeper and farther from the cost, one needs to keep in mind different kinds of problems. These plants, especially those at prototype stage, often need repair and maintenance and, therefore easy access to the power plant should be one important consideration. Also sub-marine cables for power transmission to the shore need to be well-constructed. Construction in depth increases the requirement of initial investment, which is also one of the prime considerations. In the light of the above aspects, it is often preferred to build more power plants in coast areas rather than choosing a location deep into the open sea.

6.4 Wave Power Plants

Wave power plants utilize the continuous wave movement on the water surface caused by tidal energy and wind is used by wave power plants. To be more accurate, wave energy is predominantly an indirect form of wind energy that causes movement on the surface of stretches of water. There are four different principles to convert wave power into electric power:

Fig. 6.4 LIMPET wave power plant (photograph by courtesy of Voith Hydro Wavegen)

- Wave energy turned into air pressure
- Wave energy converted into potential energy
- Different altitudes of water surface turned into mechanic energy
- Waves activate oscillating movement of a hydraulic flap

The first prototypes of power stations which converted wave energy to pressed air was built in Norway in the 1970s and in Trivandrum in South India (see also Fig. 1.8 right scheme) in the 1980s. Both prototypes operated some years successfully, but are out of operation now. The wave water was running into a case, which was closed by the coast and which was open on the seaside. A lip inside the case splits the volume in two sub volumes. When water was in the sub volume behind the lip the air inside was pressured. The pressured air drove an air propeller that turned the generator. The wave was reflected on the back wall of the case and when water ran out of the sub volume behind the lip air was sucked from the outside into the case. Controlling valves took care that air was going always in the same direction over the wind propeller.

The commercially constructed and operated power plant *LIMPET* (Land Installed Marine Powered Energy Transformer) in Scotland is based upon the principle of converting wave energy into compressed air (see Fig. 6.4). It has an overall capacity of 500 kW. The power plant is positioned in a crevice where a slanting part of the roof reaches into the water, while the other part of the roof is straight above the land. In the land-facing wall there are two wind turbines converting wind into electricity (see Fig. 6.5). The use of energy is based on the principle of the communicating vessels, which means that the water levels inside and outside of the building will try to align with one another. When the level outside is higher than inside, due to wave movement, water will enter the reservoir

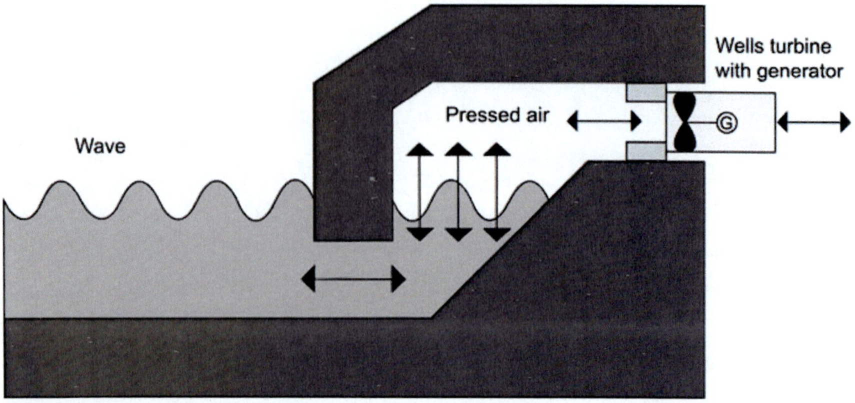

Fig. 6.5 Scheme of the LIMPET power plant (concept of the figure taken by courtesy of Voith Hydro Wavegen)

and the air inside the building will be compressed. This compressed air drives the turbine and produces electricity. As soon as the water level outside the slanting roof sinks again, the water level inside will sink as well, and air comes inside through the turbine. So called Wells-turbines are installed in the *LIMPET* power plant (500 kW rated power). They allow air to go in both directions through the turbine by rotating in the same direction.

Here, water power is not used directly, but indirectly. Actually wind energy is used here that results from the different levels of the water caused by wave movements.

The second principle of operation of wave power plants is through conversion of wave energy into potential energy. Wave energy is actually kinetic energy which needs to be converted into potential energy by lifting water up onto a higher level. This is done in the first wave power plant named *Tapchan*, which stands for *tapered channel*. It was constructed in 1984 in Norway. In this plant, waves enter a raised and tapered channel, in which the cross section of the water flow channel for inflow is narrowed. Then the water is accumulated in a reservoir from where it exits in a controlled manner through a simple propeller turbine in a pipe back into the ocean. The channel of the Tapchan plant has a 60 m wide opening through which the water flows in. The reservoir is about seven metres deep and the surface of the backmost part of the reservoir is about three metres higher than the sea level. The plant has a capacity to generate 350 kW electric power using wave energy and the average annual produced energy is nearly 2 GWh. The Tapchan power station is already out of operation.

A similar power plant that converts wave energy into potential energy is the *Wave Dragon*. In contrast to the Tapchan, which was a stationary plant at the coast, the Wave Dragon is a floating construction. In 2003, it was installed in Denmark on the coast of Nissum Bredning (see Fig. 6.6).

Another idea to convert wave energy into electricity is the Scottish experimental machine *Pelamis*, situated in Portugal. Pelamis is species of water-living snake that gave the inspiration for the name of the snake-shaped construction of

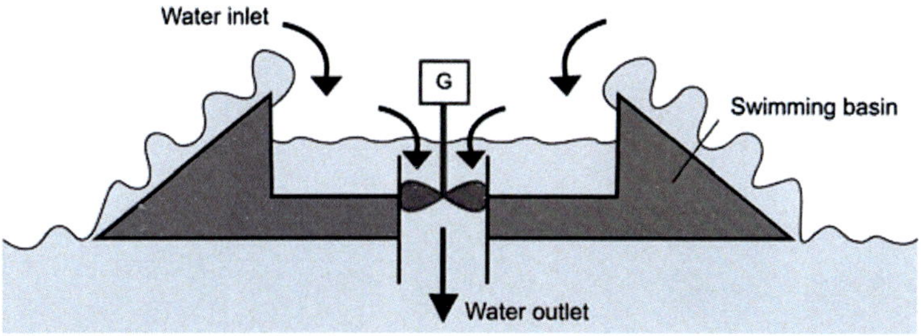

Fig. 6.6 Scheme of the Wave Dragon experimental power plant

this plant. The three, 140 m long, metal tubes constitute the main portion of this plant that started operating in 2006. It is the third principle of using wave energy that is used in this plant: the conversion of the different altitudes of the water surface into mechanic energy and finally into electricity. Each of the tubes consists of several hollow cylinders which are connected to the hydraulic pump by joints. When waves move the snake-shaped power plant in different spatial orientations the hydraulic cylinders pump the oil from the high pressure oil pump into the pressure tanks from which the oil is unable to flow back. Here the pressure is accumulated and the compressed oil exits the pressure tank through an expander. The expander converts the pressure into mechanical rotation energy which powers an electric generator while reducing the oil's pressure and leading it back to the high pressure oil pump and the process begins all over again. As is the case with all offshore installations, a submarine cable transports the energy from the ocean to the land and into the electricity system. The three tubes can deliver a maximum of 2.25 MW power, 750 kW each.

In Denmark another first of its kind wave power plant named *Wave Star* was constructed in 2006 on the northwest coast. The construction of this plant consists of a body with floats attached on either side. These floats are shaped like half balloons and are one metre in diameter (see Fig. 6.7). They float on the water surface and transmit the movement to the holder, where the motion is converted into electric power. The Wave Star starts generating energy from a wave height difference of five centimetres and its power generation capacity is up to 5.5 kW. The principle, however, which the power plant is based on, is the same as in Pelamis as the changing water levels drive oil pumps, too.

The fourth principle of utilizing wave energy which also was described in Chap. 1 (see also Fig. 1.8 left scheme) is realised by a special type of wave power plant in which the waves enter a chamber. In the chamber there is a flap. When entering the chamber the water activates the flap by pushing down its frontside, reflects on the back of the chamber and flows back. This movement pushes down the backside of the flap and the water flows back into the ocean. The movement of the flap caused by the water is transmitted to hydraulic cylinders. These cylinders pump oil into pressure tanks. The operation is similar to the of the Pelamis plant explained above. One prototype of this kind existed in Yagishiri, in Japan.

Fig. 6.7 Wave Star (photograph by courtesy of Wave Star Energy)

It consisted of three units, with a total of 450 kW electric power. Every unit was a case with of about 13 m length, 8 m width and 10 m height. The use of wave energy has to face various problems. Wave power plants are exposed to hard weather conditions, and some experimental plants have already been destroyed by the forces of tides and strong storms. Maintenance and repair are costly, and there are still few experience data. This makes financing of wave power plants even more difficult as banks hesitate to give credits for insecure projects.

6.5 Ocean Thermal Power Plants

Water bodies have a vertical thermal gradient. The water on the surface and subsurface of the ocean is warmer than the water in deeper layers. The difference in temperature in different depths can also be converted directly into electricity by an OTEC (*Ocean Thermal Energy Conversion*). As per the rule of thumb, after every 100 m depth, the temperature of water changes by 5 K. In a typical OTEC power plant as shown in Fig. 6.8, there is a closed circulation system that contains a fluid working medium such as an organic fluid or ammonia, which is moved from deeper layers of the system into shallow layers and back into the deeper ones again. The choice of working medium depends on the temperature of water in the

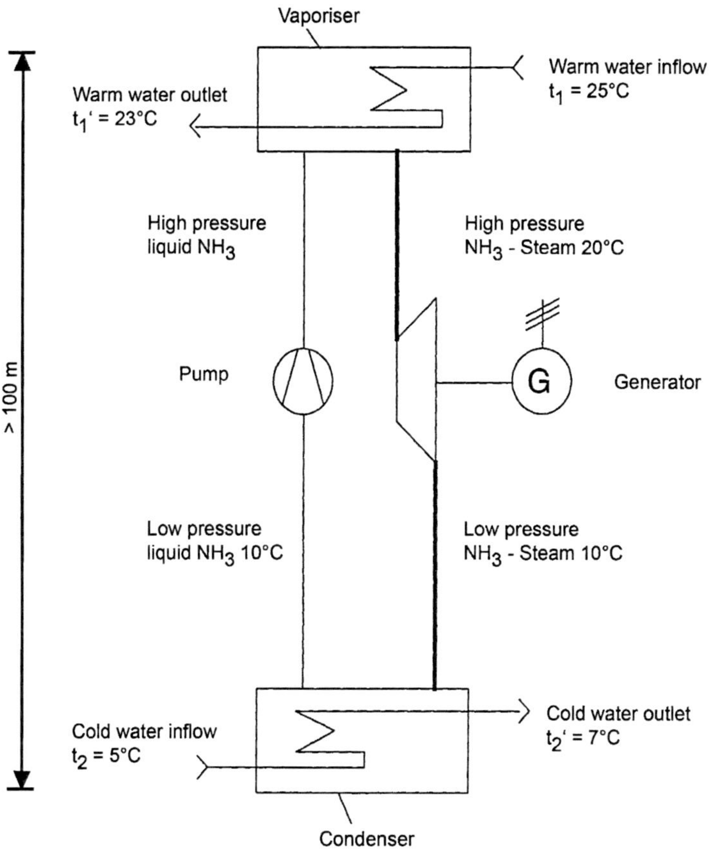

Fig. 6.8 Scheme of an oceanic thermal power plant OTEC

shallow and deeper layers of water. The circulation of working fluid is not set off by itself, therefore a pressure pump is required to pump the medium from deeper levels to the upper levels that simultaneously also generates a higher pressure in the upper part of the system. Here, in the upper part, the medium reaches a vaporiser where it gets evaporated. For example, the 25 °C warm surface water enters the vaporiser, and transmits warmth from ocean water to the working fluid for its evaporation. Therefore while exiting the vaporiser, the water outlet temperature is a few degrees cooler. The delivered energy helps the working medium to evaporate. The vapour enters a steam turbine which drives a generator. In the turbine the pressure and the temperature of the vapour is reduced. Leaving the turbine the vapour will flow to the condenser. It is located at the deepest part of the power plant, maybe 100 m and more. The incoming ocean water has, for example, a temperature of 5 °C, condenses the vapour and is a few degrees warmer when exiting. Vapour is now converted into liquid again and it is pumped with high pressure in the system for restart the cycle.

In short, in the operation of an OTEC power plant, the cool liquid medium streams upward from the condenser, and the pressure is low. The pump increases the pressure. Then the medium reaches the vaporiser where it gets evaporated by receiving heat of warm water. The warm, evaporated working medium, that has a higher pressure, drives the turbine where pressure and temperature are reduced. The turbine drives the generator. Then, finally, the medium reaches the condenser where it turns into a liquid state again. The process is generally the same like in every steam power station, only the working medium is different from water and steam.

Similar to other offshore power plants, this plant also requires a submarine cable for the transmission of generated power from the ocean to the land. The efficiency factor of an ocean thermal power plant is of the order of three percent which remains one of the major bottlenecks for its commercial scale use.

The advantages of ocean thermal power plants are certainly the emission-free power generation. Also, the power plants do not take up all the room of the oceans' surface and do not need much space. The availability of energy is also relatively non-fluctuating and secure. The most important advantage of these systems is that ocean water does not cool down quickly after the sun has set, which results in the possibility of power generation in the late evening hours after sunset, just at the time when many households demand energy and additional power is required by consumers.

On the other hand, ocean thermal power plants have not yet been developed to an economical level. There are some practical problems involved with their operation. Limpets, for instance, and other molluscs that get attached to the power plant water openings, restrict the temperature-dependent circulation and reduce the efficiency of power generation. Another problem is that these power plants are not very well accessible for maintenance and repair as they are not very close to the land and stretch deep into the water. Although they are anchored in the sea bottom, mighty storms may harm the machines. These aspects involve complications with the operation of ocean warmth projects. Few investors are willing to give sufficient funds to expensive experiments and construction. The efficiency is not very high, as 80–90 percent of the produced energy is needed for the pumping. An experimental plant was constructed in Japan in the year 1981. This plant has a power generation capacity of 100 kW, 90 percent of which was consumed by the pressure pump used to run the system. Another ocean thermal power plant worked successfully in the years 1993–1998 in Keahole Point in Hawaii. In late summers, when the water surface was particularly warm, the plant generated 250 kW, out of which 200 kW were needed for the pressure pump. Experience is still limited so that further research and development is needed to improve this type of power plant.

6.6 Osmotic Power Plants

The osmotic power plant uses the energy of, as the name suggests, *osmotic power* or the energy available due to *salinity gradients*. Osmosis is the process of directed movement of molecules through a selective membrane. In an ideal

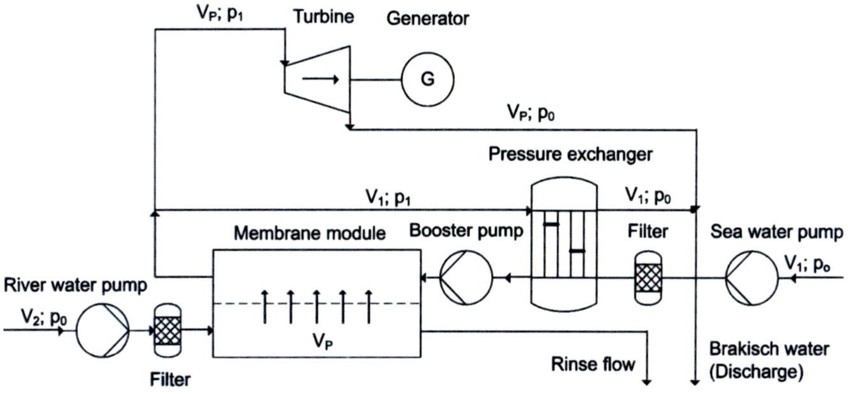

Fig. 6.9 Scheme of an osmotic power plant

selective membrane, molecules of a certain type are able to pass through the membrane in both directions, e.g. water molecules, while other components, like ions (from salt molecules), are unable to cross the membrane. Due to this directed selective diffusion, a certain pressure gradient is created, the so-called *osmotic pressure*.

When salt water and fresh water come into contact the mixture has the natural tendency to equalize the concentrations. In case of an attempt of mixing salt and fresh water over a selective membrane, the result is a net volume flow of water molecules from the fresh water side to the salt water side of the membrane, by diluting the salt water (principle of osmosis). The process stops as soon as osmotic pressure and hydrostatic pressure, i.e., the pressure of water evoked by gravity, are equally high on both sides of the membrane. At this point a thermodynamic equilibrium has been reached.

Figure 6.9 shows the scheme of an osmotic power plant. Based on the principle of osmosis, the difference in salinity between sea water and fresh water can be used for generating power.

Sea water and river water need to be separated in order to avoid a mixture of both liquids before they enter the power plant. Therefore a distance between the sea water and the river water intakes is required. Suitable locations for osmotic power plants can be found at the mouth of rivers where fresh water from rivers is about to get mixed with salt water from the sea.

The main components of osmotic power plants are: entrance and exit for sea water and river water, pumps, pre-treatment (e.g. filters), pipes, membrane modules, and a pressure exchanger system for highly efficient energy recovery.

The technical process of energy conversion based on osmosis is called pressure-retarded-osmosis (PRO). In PRO the sea water $\left(\dot{V}_1\right)$ is pre-pressurised to typically half the osmotic pressure, before the sea $\left(\dot{V}_1\right)$ and the river water $\left(\dot{V}_2\right)$ are brought us with each other in the membrane modules. It is the osmotically driven increase

Fig. 6.10 Osmotic power plant prototype in Hurum/Norway, opened November 2009 (photograph by courtesy of flickr (Statkraft))

of pressure of the permeate volume flow $\left(\dot{V}_P\right)$ across the membranes that is used in osmotic power plants to drive a water turbine as shown in the figure.

The mixing of sea water and river water is part of the natural global water cycle. The sun evaporates surface sea water, thus increasing the salinity of the top layer from which evaporation takes place. The evaporated water returns as rain and flows into the rivers, while the salt remains in the sea water. The salinity gradient between sea and river water at any location is therefore dependent upon the intensity of solar radiation. One main advantage of osmotic power plants is the opportunity for non-fluctuating power generation, comparable to conventional hydropower plants.

The average salinity of the world's ocean water is 3.5% which results in an osmotic pressure of up to 27 bar. This pressure is equivalent to a water head of 270 m, hence there exists a power generation potential. However, in PRO approximately only half of the pressure is used for energetic gains, because the power plant is typically operated at half the osmotic pressure (pressure of the incoming sea water). This is due to characteristics of the membrane power, which reaches a maximum at these operating conditions. Under these conditions a volume flow of 1 m^3/s of river water facilitates an installed capacity of around 0.6–0.8 MW for an osmotic power plant. The volume flow ratio between river and sea water is typically in the range between 1 and 2.

Plans and theories for osmotic power plants are quite old but the commercial use of this concept was restricted due to the non availability of efficient membranes at ancient time. The first osmotic power plant was constructed in Norway (Fig. 6.10) with an installed power of about 4 kW. It has a membrane area of about 2.000 m^2. The aim of the prototype is to test the specially

Table 6.1 Status of tidal power plants

Name	Location	Power	Year of construction
Sihwa	South Korea	260 MW	Under construction
La Rance	France	240 MW	1966
Kislogubsk	Russia	0.4 MW	1968
Jiangxia	China	3.2 MW	1980
Annnapolis Royal	Canada	18 MW	1984

Table 6.2 Status of ocean current plants

Name	Location	Power	Year of construction
KOBOLD	Italy	40 kW	2001
Stingray	Great Britain	150 kW	2002
Seaflow	Great Britain	350 kW	Out of operation
Seagen	Great Britain	1.2 MW	2010

designed PRO-membranes under conditions of daily operation. The project is mainly supported by the EU, the Norwegian company Statkraft, and the country of Norway.

On a global basis the technical potential for power generation by osmotic power plants is estimated to be approximately 5,200 TWh per year and for Europe to be approximately 400 TWh per year. The ecological potential takes additionally to the technical constraints of the energy conversion, the ecological restrictions of water extraction into account. That value is significantly lower than the technical potential. For details see [1].

For future development the main task is to continue the improvement of membranes (membrane power, salt rejection etc.) as the currently available membranes are not good enough yet for the operation of a commercial power plant.

6.7 Survey of Ocean Energy Facilities

Tables 6.1, 6.2, 6.3, 6.4, and 6.5 give a survey of all ocean energy facilities which have been constructed and tested. A lot of them are out of operation in our days. The information was collected by the authors as thoroughly as possible and is representative of the status of knowledge at the end of the year 2010.

Despite the fact that most of technologies related to ocean energy are still not nature enough for commercialization, since they offer non polluting and

Table 6.3 Status of wave power plants

Name	Location	Power	Year of construction
Tapchan	Norway	350 kW	Out of operation
Toffelstallen	Norway	350 kW	1986
Islay	Great Britain	75 kW	1987
Vizhingam	India	150 kW	Out of operation
Wave Power	Japan	30 kW	1987
Mighty Whale	Japan	110 kW	1998
Sakata Port	Japan	60 kW	1989
Dawanshan	China	20 kW	1990
Islay	Great Britain	500 kW	2000
Wave Dragon	Denmark	20 kW	2003
Port Kembla	Australia	300 kW	2004
Pelamis	Great Britain	750 kW	2004
Wave Star	Denmark	1,8 kW	2006
Pelamis	Portugal	3 × 750 kW	2008

Table 6.4 Status of ocean thermal energy plants (OTEC)

Name	Location	Power	Year of construction
Nauru	Japan	100 kW	1980
Keahole Point	USA/Hawai	210 kW	1993–1998
Saga University	Japan	75 kW	1985

All plants are out of operation

Table 6.5 Status of osmotic power plants

Name	Location	Power	Year of construction
Hurum	Norway	4 kW	2009

non-depleting source of power, more research and development needs to be done in these fields.

Reference

1. Stenzel P, Wagner HJ (2010) Osmotic power plants: potential analysis and site criteria. 3rd international conference on ocean energy, Bilbao, 6 Oct 2010

Chapter 7
Economics of Hydropower Plants

7.1 Cost and Benefits

The economics of a hydropower plant is quite different from that of any other type of power plant since various considerations such as water supply, irrigation and river navigation are involved besides regular economic aspects of cost of generated power. In fact, some of these aspects, such as effect on irrigation or recreation facilities are difficult to quantify. Hence, true economic analysis of a hydro power plant, especially a large hydro power plant, is a mix of quantitative and qualitative approaches.

The major benefits and cost components for estimating annual the net benefits from a hydropower plant are shown in Fig. 7.1 and defined below:

1. Gross power benefits: These benefits reflect the income from sale of power or avoided cost of power if the hydropower plant did not existing and power has taken from costlier source.
2. Benefits of avoided pollution: Relative to alternative types of power generation, such as a coal-fired plant, hydropower production generates less air pollution or greenhouse gases. The avoided pollution is considered as a benefit of hydropower projects.
3. Costs of operation: This type of costs reflects investment costs for the project, anticipated future reinvestment costs, and current operation and maintenance (O&M) costs.
4. Benefits of project services: Beyond power generation, hydroelectric projects may offer benefits such as flood control, water supply, irrigation, river navigability and improvement of infrastructure and economical prosperity of the region.
5. Costs of environmental measures: Many licensing decisions introduce operating conditions designed to protect, mitigate damages to, or improve environmental quality. These changes may result in direct costs and/or reduced power

H.-J. Wagner and J. Mathur, *Introduction to Hydro Energy Systems*, Green Energy and Technology, DOI: 10.1007/978-3-642-20709-9_7,

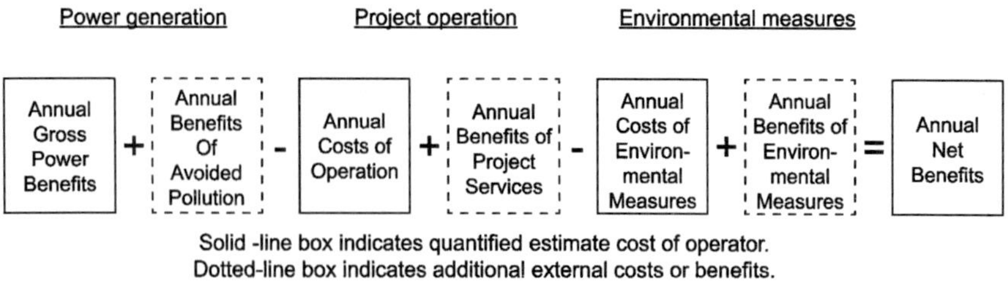

Fig. 7.1 Overview of economic analysis

values from the viewpoint of the hydropower station owner. There are direct costs associated with, for example, construction of fish passage facilities. Similarly, due to environmental measures to protect flora and fauna sometimes flow of water is restricted that may reduce power generation either because they cause direct losses in available flow or they shift power generation from periods when energy prices are high to periods when energy prices are low.

6. Benefits of environmental measures: Environmental measures, such as fish screens or changes in minimum flow requirements, can improve fish and wildlife resources, recreational opportunities, and other aspects of environmental quality. Since these benefits are different from the direct revenue from sale of power, they are often referred to as "non-power" benefits.

Details related to the cost structure and sale of power are presented in the following sections.

7.2 Cost Structure

7.2.1 Initial Cost

The cost of hydropower plants can hardly be generalized since every site may offer a unique set of challenges, such as the lengths of pipes and tunnels, the difficulty in transporting equipment due to a poor road network, or necessary investments in infrastructure, different geology etc. that are reflected by the plant cost and hence on the cost of generated power. The initial costs of hydropower plants are usually found to vary between 1,000–5,000 Euro per kW depending on the size of country and location. In the case of the Itaipu hydropower plant, 30 years ago the capital cost was around 1,300 Euro per kW (on the basics of the US$-Euro relationship at the start of 2011), whereas in case of the Three Gorges Dam plant it was, according to press releases 1,000 Euro per kW. When relatively smaller capacities are planned, the capital cost per kW is higher than these values.

In case of green field projects, where no dam has existed before the hydropower project, civil engineering works typically account for 65–75% of the total initial cost, and meeting the environmental and legal criteria requires 15–20%. The cost of plan machinery such as turbine, generator and control systems may account for only 10% of the total initial cost. There are, of course, no fuel costs associated with hydro power plant during operation; hence operation and maintenance costs are quite low in comparison to a fossil power station.

In case of non green field projects, i.e. in cases where dams already exist for some other purposes such as recreation or flood control; the locations can be upgraded to have hydro power plants at relatively lower costs. Similarly, run-of-river could also be less costly as compared than reservoir type power plants, as less civil works are involved.

7.2.2 Operation Cost

The operation and maintenance cost of hydropower plants comprises three major components: Maintenance of plant machinery including replacement of part, salary of staff, and insurance. The life of hydropower plants ranges from 20 years to 40 years and beyond. For financial calculations, usually a calculative lifetime of around 30 years is considered.

7.2.2.1 Maintenance Costs

The maintenance cost of plant machinery has two major components, preventive maintenance and breakdown maintenance. Per year, approximately 3–5% of the capital investment can be considered as O&M cost of plant equipment in the initial years, which usually increases as the plant gets older.

The turbine runner is the part that requires most of the maintenance work. Due to the cavitation effect or due to the hammering action of the silt arriving with the water, the runner blade gets damaged. This damage is predominant in some countries during the rainy season when the incoming water in the reservoir carries lot of eroded sand, and as the retention time in the reservoir or the stilling basin is reduced, the quality of the water arriving at the turbine is relatively poor. For this reason it is general practice during the season when the power demand decreases, to remove the runners of a multi-turbine power plant for repair turn by turn.

In addition to maintenance, another important cost element is insurance costs. Insurance of hydro turbines and dams is required to secure the loss of their high investment fully or partially in cases where parts are damaged and also to cover the huge possible loss due to flooding if something goes wrong with the dam.

7.2.2.2 Plant Utilization Factor

In theoretical calculations, hydro power plants are assumed to be available for power generation whenever water is available. In practice, however, this may not be true in all situations. With modern preventive and predictive maintenance practices of advanced plant control, the availability of hydropower plants has increased over the past decades. The best plants may be available for about 95% of the time. The unavailability of water for full capacity utilization of a hydropower plant though becomes a constraint. Due to this factor, the plant utilization ranges from 60–80%. In some very good locations, where water availability is consistent due to a mix of rain-fed and snow melting systems, and where the reservoir has a large capacity of storage, an even higher plant utilization factor (PLF) can be achieved. On the other hand, in case of very small plants, this factor may even be lower than 50%.

7.2.2.3 Salary and Administrative Expenses

Since a hydropower plant requires continuous monitoring and maintenance, the salary component cannot be ignored in the financial analysis. Old power plants used to require more persons to operate and control various systems, whereas, due to automatic controls, new plants require much less manpower.

7.3 Electrical Tariffs

Different types of tariff systems are found in the case of hydropower plants due to differences in policies in different countries/locations all around the world, as described below. Important influences on the decision to introduce a certain tariff system are liberalized or non liberalized markets, plentiful or shortage of electricity available, incentives for clean electricity production etc.

7.3.1 Feed-in Tariff

The feed-in tariff scheme, as its name suggests is based upon the principle of paying amount to the power producer as per the amount of electricity fed into the grid. This is done at a pre-declared rate per unit of electricity. This rate is than the rate of production of electricity from a conventional (using fossil or nuclear fuels) power plant. The most important aspect of a feed-in tariff system is that the grid operator cannot deny the acceptance of the power generated by the hydropower plant even if it is surplus. In this case other non hydropower stations must reduce their power generation. The feed-in tariff system exists e.g. in Germany, India as well as in many other countries.

In Germany every operator of a hydroelectric power station received an amount between 3.5 Euro cent/kWh and 12.7 Euro cent/kWh (as per October 2010), depending of the capacity and the age of the power station.

7.3.2 Availability Based Tariff (ABT) System

ABT is a mechanism for the recovery of fixed charges of a power plant or transmission licensee through the commercial means of incentives or disincentives. It is a performance-based tariff for the supply of electricity by generators owned and controlled by the central government or those which are involved in selling power in more than one state. It is also a new system of scheduling and dispatching power, which requires both generators and beneficiaries to commit to day-ahead schedules. It is a system of rewards and penalties seeking to enforce day ahead pre-committed schedules, though variations are permitted if notified one and one half hours in advance. It facilitates grid discipline and helps in trading capacity and energy and facilitates the merit order dispatch of power.

Usually, the ABT has three parts: A fixed charge (FC) payable every month by each beneficiary to the generator for making capacity available for use. The fixed charge may not the same for each beneficiary. It usually varies with the share of a beneficiary in the overall power generation capacity. The fixed charge, payable by each beneficiary, also varies with the level of availability achieved by any generator.

The second part is the energy charge payable for every kWh of energy supplied as per a pre-committed schedule of supply drawn upon a daily basis.

The third part of ABT is a charge for unscheduled interchange (UI charge) for the supply and consumption of energy in deviation from the pre-committed daily schedule. This charge varies inversely with the system frequency prevailing at the time of supply/consumption.

7.3.3 Bulk Electricity Tariff System

Bulk tariff (for central generating company or the generating company which sells power to more than one state) means the annual fixed charges (AFC) in respect of each hydropower station which is determined by the Central Electricity Regulatory Commission. The components of AFC calculation are:

1. Interest on loan capital
2. Depreciation.
3. Return on equity.
4. Operation and maintenance expenses.
5. Interest on working capital.

The AFC is recovered in the form of capacity charges (50% of AFC) and energy charges (50% of AFC).

7.3.4 Time Dependent Rates

In some regions state owned electricity companies are more interested in buying electricity during the periods of peak load time at maximum consumption on the electrical grid, because in this way they may save using the electricity from the less efficient generating units to produce power at higher costs. Therefore, in some areas, power companies apply variable electricity tariffs depending on the time of day, when they buy electrical energy from private power plant owners. Normally, power plant owners receive less than the normal consumer price of electricity, since that price usually includes payment for the power company's operation and maintenance of the electrical grid, plus its profits. In case of location where feed-in tariff system exists as in Germany, the time of generation of power becomes non-significant.

7.3.5 Quota System or Renewable Energy Certificates

One more system is that of allocation of 'quota' of renewable energy. In this system, every producer of electricity for grid is given a 'quota' e.g. 20%. The total electricity produced by every company has to include a 20% share of energy coming from renewable sources such has hydro, wind and solar. If this quota is not met provisions for penalties have been taken. The European Union discussed adopting this system in Europe. Provisions are being developed to provide that if any company produces energy from renewable in excess of its quota, e.g. 25%, it would be granted a certificate for this excess renewable energy, called Renewable Energy Certificate. This certificate can be purchased by other companies with a smaller share of renewables than the quota of e.g. 15%. With the purchase of certificates, the second company will also be considered to be complying with its quota. This system is quite similar to the concept of 'emission trading' for greenhouse gases. Up to now, no consensus had been reached for introducing such a quota system in Europe. There was a debate on the argument that with such quota system, relatively costlier renewable energy technologies such as solar photovoltaic systems would not be preferred at all. As a consequence the technological progress on these technologies would be stopped. In India, the Renewable Energy Certificates (REC) were launched as late as 2010. It is based upon the philosophy of a quota system or Clean Development Mechanism (CDM). Every state has been given a target of producing a minimum percentage of renewable energy. In case a state is not able to achieve the target, it can buy Renewable Energy Certificates from other producers to meet its deficit of renewable energy share. This mechanism works similar to that of a stock exchange.

7.3.6 Production Tax Incentives/Investment Incentives

The Production tax incentive is a generation-based mechanism, which supports renewable energies through payment exemptions from electricity taxes, e.g. the energy tax for renewable energies, applied to all producers. Hence it is a system that affords an avoided cost on the producer side. Also the investment incentive is a mechanism, as the name implies for lowering the costs for the investment in renewable energies so that it gets more attractive funding.

7.3.7 Environmental Credit and Clean Development Mechanism

Many governments and power companies around the world wish to promote the use of renewable energy sources. Therefore, in developed or industrial countries, they offer a certain environmental premium to renewable energy coming from hydro, solar or wind power plants, e.g. in the form of refund of electricity taxes etc. on top of normal rates paid for electricity delivered to the grid. In developing countries like India, due to the fact that every unit of electricity generated avoids generation of the same amount of electricity in fossil fuel based power plants, the avoided environmental emissions offer an opportunity of getting for additional earnings through the Clean Development Mechanism under the Kyoto Protocol. Every ton of carbon dioxide gas saved or avoided, termed as one carbon credit, was sold to industrialized countries at rates of about 15 Euro in the year 2010.

Chapter 8
Outlook for Hydropower

By the year 2050, the world population is expected to have increased by 50%, as compared to its level in the year 2000. The energy consumption per inhabitant per year is usually governed by the standard of living of the population. Today the less developed countries in the world, with 2.2 billion inhabitants, have an annual per capita consumption of primary energy which is 20 times lower than that of the industrialized countries with about 1.3 billion inhabitants. Per capita electricity consumption of less developed countries is about 35 times lower than the consumption in industrialized countries. It is, therefore, certain that world energy consumption, and especially electricity consumption, will increase considerably during this century. Thus, the challenge can be clearly defined: an inevitable increase in energy consumption in the world, with the risk of a major environmental impact, and climate change, as a result of the combustion of fossil fuels. The right for development is a basic human right, and there is no possible development without energy supply. In view of this situation, all available sources of energy will be necessary, but for environmental reasons, the first priority should be the development of all the technically, economically and environmentally feasible potential from clean, renewable energy sources, such as hydropower.

Concerns over disruptive fossil fuel markets and uncertain pricing, the current decline in nuclear energy as a viable energy source and the significant environmental consequences of thermal energy sources increased the emphasis on sustainable energy policies over the past few decades that include the significant development of renewable energy supplies. Common thinking often relates renewable energy to electricity production from systems utilizing wind energy, solar energy, biomass, geothermal energy and similar sources. However, the largest source of renewable energy comes from a much proven and time tested technology: hydropower stations. In addition to offering a large potential for economic power generation, unlike most renewable energy based technologies, it also provides an option to store energy without conversion from its naturally available form.

H.-J. Wagner and J. Mathur, *Introduction to Hydro Energy Systems*, Green Energy and Technology, DOI: 10.1007/978-3-642-20709-9_8,

The world's total technically feasible hydropower potential is estimated to be around 14,000 TWh/year, of which about 8,000 TWh/year is considered economically feasible. Currently, hydropower plants supply about 20% of the world's total electricity. Interestingly, hydro supplies more than 50% of national electricity in about 65 countries, more than 80% in 32 countries and almost the entire electrical power supply in 13 countries. A number of developing countries, such as China, India, Iran and Turkey, have large-scale hydro power development programs.

When assessing future energy production, clearly the policies in favor are those which emphasize sustainability and the maximum use of renewable energy to meet future needs. Consequently, one cannot afford to dismiss any form of renewable energy from the supply mix. While it is acknowledged that hydropower has significant positive and negative consequences for society and the environment, it is also recognized that all forms of infrastructural development, and in particular energy development, have different degrees of impacts.

The International Hydropower Association (IHA) [1], the Implementing Agreement on Hydropower Technologies and Programmes of the International Energy Agency (IEA) [2], and the International Commission Large Dams (ICOLD) [3], are among the leading world-wide organizations that are proponents of responsible hydropower development.

As per a study conducted by the IEA/Hydropower Agreement on Hydropower and the Environment, the hydropower policy for any country should be based upon the analysis of virtually all environmental aspects of hydropower addressing the issues of hydropower development while offering reasonable solutions for future development. Important considerations of social, cultural and economic impacts, as well as impacts on the natural environment play a vital role in considering the potential ramifications of development. As per the approach suggested by IEA/Hydropower Agreement following points must be considered by all countries:

- The need for an energy policy framework: Nations should develop energy policies which clearly set out rational objectives regarding the development of all power generation options, including hydropower, other renewable sources, and conservation.
- A decision making process: Stakeholders should establish an equitable, credible and effective environmental assessment process which considers the interests of people and the environment within a predictable and reasonable schedule. The process should focus on achieving the highest quality of decisions in a reasonable period of time.
- Comparison of hydropower alternatives: Project designers should apply environmental and social criteria when comparing project alternatives, to eliminate unacceptable alternatives early in the planning process.
- Improving environmental management of hydropower plants: Project design and operation should be optimized by ensuring the proper management of environmental and social issues throughout the project operation cycle.

- Sharing local benefits with local communities: Local communities should benefit from a project, both in the short term and in the long term. Together, these five categories of recommendations constitute a sustainable approach to renewable hydropower resource development.

It is expected that during the next few decades, there will be a multi fold increase in hydro power generation worldwide. On one hand, large power plants such as Three Gorges Dam, China and Itaipu, Brazil have set new standards while offering advantages of low electricity generation cost together with offering economic benefits related to irrigation, flood control, tourism, fishery and other developmental activities. On the other side, micro hydro power plants are also turning out to be extremely useful in remote locations of developing and under-developed countries.

Technological advancements related to the operation and control of power plants, corrosion resistance of turbine blades, dam safety, and integration of pumped storage power plants, are improving the reliability and availability of power plants. Oceanic power plants, however, have to undergo many improvements in the years to come and significant research is still required to make them economically viable.

Nevertheless, it can be concluded that with harnessing most of the hydropower available in various forms, the energy and environmental problems of the world can be solved to a large extent without compromising on economical growth and developments worldwide.

References

1. International Hydropower Association (IHA) http://www.hydropower.org
2. International Energy Agency (IEA/Hydro) http://www.iea.org/index.asp
3. International Commission Large Dams (ICOLD) http://www.icold-cigb.net/

About the Authors

Prof. Dr.-Ing. Hermann-Josef Wagner holds postgraduate and doctorate degree in electrical and mechanical engineering from the Technical University of Aachen, Germany. He is Professor for Energy Systems and Energy Economics at the Ruhr-University of Bochum, Germany. He worked as a scientist for the Juelich Research Centre, for the German Parliament and for different universities. His relevant experiences are in the fields of energy systems analysis, renewable energies like wind and water energy and life cycle analysis. Email: lee@lee.rub.de

Dr.-Ing. Jyotirmay Mathur is a mechanical engineer with a postgraduate degree in energy studies from Indian Institute of Technology, Delhi, India; and doctorate from the University of Essen, Germany. He specializes in the areas of renewable energy systems, energy policy modeling and energy efficiency in buildings. As Associate Professor in the Mechanical Engineering Department of Malaviya National Institute of Technology, Jaipur, Dr. Mathur has been the founder coordinator of the postgraduate program in energy engineering and is presently working as coordinator of the Center for Energy and Environment at his institute. Email: jyotirmay.mathur@gmail.com; jyotirmay@mnit.ac.in

Glossary

Alternating Current (AC) An electric current that reverses its direction at regularly recurring intervals, usually 50 or 60 times per second. Today's grids are operated by Alternating Current.

Bar Unit for water pressure, 1 bar is equal to 0, 1 Megapascal (MPa). One per mille bar is called 1mbar. One bar is the pressure of a water pile of 9.81 m height.

Gigawatt (GW) A unit of power equal to 1 billion Watts, 1 million kilowatts, or 1,000 megawatts.

Joule (J) A standard international unit of energy; 1 J is equal to $\frac{1\text{kg} \cdot \text{m}^2}{\text{s}^2}$ in the SI-unit system and equal to 1 Watt second (Ws), 1,055 Joules are equal to 1 BTU.

Kelvin (K) International unit of temperature. Zero Kelvin is equal to 273.15°C. Kelvin is also used as unit for temperature differences.

Kilowatt (kW) A standard unit of electrical power equal to 1000 watts, or to the energy consumption at a rate of 1,000 joules per second.

Kilowatt hour (kWh) 1,000 thousand watts acting over a period of 1 hour. The kWh is a unit of energy. 1 kWh = 3,600 kJ.

Megawatt (MW) 1,000 kilowatts, or 1 million watts, standard measure of electric power plant generating capacity.

Megawatt hour (MWh) 1,000 kilowatts hours or 1 million watt hours.

Watt (W) The rate of energy transfer equivalent to one ampere under an electrical pressure of one volt. One watt equals 1/746 horsepower, or one joule per second. It is the product of voltage and current (amperage).

Watt hour (Wh) A standard international unit of electrical energy. 1 watt acting over a period of 1 hour is equal to 1 Wh.

Index